Oklahoma Horizons Series

Oklahoma Horizons Series

THE RUSH BEGINS

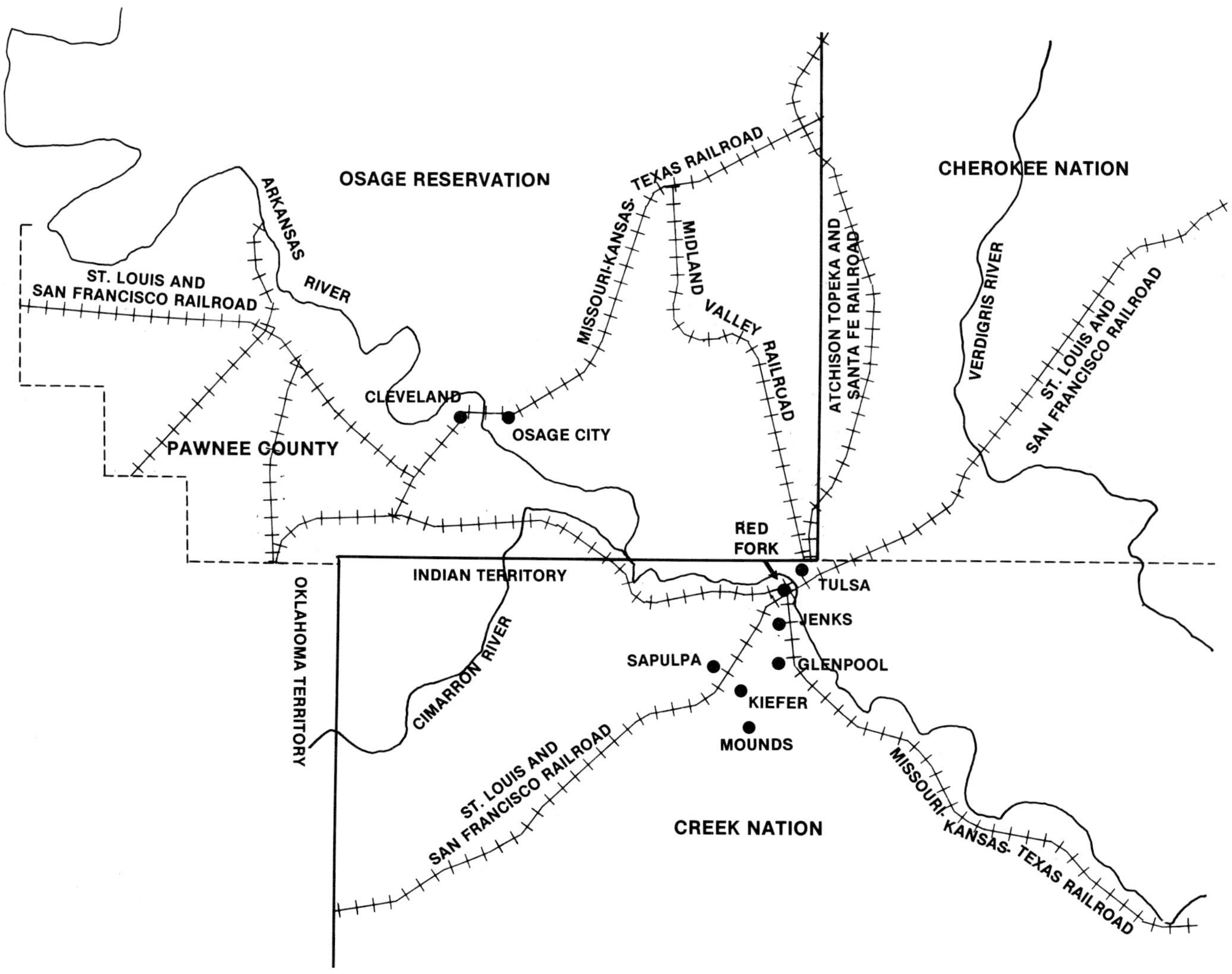

The area where the Twin Territories (Oklahoma and Indian) joined along the Arkansas River. Note that Red Fork and Cleveland are located in bends of the river. According to the theory of "creekology," streams flow around underground formations containing oil and therefore the bend of a stream is the best place to drill. The interlocking grid of railroads easily can be seen. The availability of a means for transportating crude to market played a major role in the early development of Red Fork, Cleveland, and Glenn Pool. Also note that the three fields are within 40 miles of one another. Despite their proximity, the development of Red Fork and Glenn Pool, which were in the Creek Nation, was hampered by federal regulations governing drilling operations on Indian land, whereas drilling at Cleveland, which was in Pawnee County, Oklahoma Territory, was allowed to proceed unimpaired. (Oklahoma Heritage Association).

Oklahoma Horizons Series

THE RUSH BEGINS:
A HISTORY OF THE RED FORK, CLEVELAND AND GLENN POOL OIL FIELDS

By
Kenny A. Franks

Published for the Oklahoma Heritage Association
by Western Heritage Books, Inc.

For Douglas William Franks

Manufactured in the United States of America for the Oklahoma Heritage Association, Oklahoma City.

ISBN: 0-86546-055-8

PREFACE AND ACKNOWLEDGEMENTS

This work is the fourth in a series of books undertaken to provide an in-depth understanding of the growth and development of the Oklahoma petroleum industry. The need for additional work became apparent as the final draft of *The Oklahoma Petroleum Industry* was being prepared. Although that work contained 284 pages of large format text, the subject was so complex that only a cursory treatment could be alloted to the major discoveries. For example, the story of the Cleveland, Red Fork, and Glenn Pool strikes was told in 20 pages. Obviously, such a limited treatment could not do justice to the huge role that the development of these three pioneering fields played in the Oklahoma petroleum industry. To fill the void, the Oklahoma Heritage Association launched a multi-volume series on the great oil discoveries of Oklahoma. These examinations allowed the preparation of detailed texts, complemented with numerous historical photographs, which offered a visual narrative of this important aspect of the state's unique history and heritage.

This book would not have been possible without the cooperation of John T. Griffin, President of the Oklahoma Heritage Association; Dr. Paul F. Lambert, Executive Director of the OHA; and the members of the Association's Horizons Committee. In addition, had it not been for the Oil and Gas Oral History Advisory Committee of the OHA, much of the material utilized in completing this work would not have been available. Its members are: W. F. Martin, Bartlesville, Chairman; Dean A. McGee, Oklahoma City, Vice-Chairman; John H. Williams, Tulsa, Vice-Chairman; Ralph Abercrombie, Tulsa; Jack Abernathy, Oklahoma City; Carl Anderson, Jr., Oklahoma City; J. W. Bates, Jr., Tulsa; Tom Boyd, Bartlesville; William B. Cleary, Oklahoma City; William H. Bell, Tulsa; Sam Cerny, Oklahoma City; H. H. "Bud" Champlin, Enid; Dr. Robert J. Collins, Jr., Oklahoma City; Lee Daniel, Tulsa; Paul Endicott, Bartlesville; and Robert A. Hefner, Jr., Oklahoma City. Other committeemen are: Robert A. Hefner, III, Oklahoma City; Walter Helmerich, III, Tulsa; T. K. Hendrick, Oklahoma City; Jack Hodges, Oklahoma City; C. C. Ingram, Tulsa; Warren Jensen, Ponca City; Jack D. Jones, Tulsa; James O. Kemm, Tulsa; John E. Kirkpatrick, Oklahoma City; P. C. Lauinger, Tulsa; P. C. Lauinger, Jr., Tulsa; James P. Linn, Oklahoma City; Jack Maurer, Duncan; Robert C. Moore, Tulsa; Walter Neustadt, Ardmore; Sam Noble, Ardmore; E. L. Paramore, Dallas; John Steiger, Tulsa; Robert E. Thomas, Tulsa; and Lew Ward, Enid.

Much of the material in the various books was taken from interviews of actual participants of the oil boom. Many of these interviews were made by the Writers Project of the Works Progress Administration in the 1930s. Although an important contribution to the preservation of oil field heritage, the WPA project was not as extensive as it could have been. To complement the WPA effort, the Oklahoma Heritage Association launched its own Oil and Gas Oral History Project specifically to preserve this unique legacy for its series of books.

The last major boom, prior to the Deep Anadarko play of the late 1970s and early 1980s, occured in the early 1930s. In the intervening half century, many of the people who lived through the oil rush died and as time passes even fewer will be available to researchers. Inasmuch as historians largely have ignored the Oklahoma oil boom until recently, with the passing of these pioneers much of the unwritten history was lost with them. It is hoped that the series on Oklahoma's great oil strikes and the adjunct interviews will preserve this legacy.

Many people, companies, and organizations helped in the research of this work. Without their cooperation it could never have been completed. Among them are: Andrea Clark, Oklahoma Historical Society; Peggy Coe, Oklahoma Department of Libraries; Ruby Cranor, History Room, Bartlesville Public Library; Karen Fite, Oklahoma Department of Libraries; Carol Guilliams, Oklahoma Department of Libraries; Mary Hardin, Oklahoma Department of Libraries; Julia Koehler, Oklahoma Department of Libaries; Kim Kirkpatrick, Oklahoma Historical Society; Fred Standley, Oklahoma Historical Society; Mary Sweeney, Oklahoma Department of Libraries; Henry J. Sydloski, Oklahoma Heritage Association; and Charlesa Timmons, Oklahoma Historical Society. Other organizations contributing photographs were the Interstate Oil Compact Commission and the Oklahoma Geological Survey.

Much of the material on Cleveland was taken from the collection of the Triangle Heritage Association of the Cherokee Outlet's Museum. Three individuals in particular, Robert L. Lumsford, III, Louise Armstrong, and Robert Owen Jordan deserve much of the credit for the preservation of Cleveland's petroleum legacy. Lumsford is the third generation of his family to be involved in the banking business in Cleveland and serves as Treasurer of the Triangle Heritage Association of the Cherokee Outlet. Mrs. Armstrong, curator of the local museum, has spent many hours collecting memorabilia of the area's heritage. Jordan, whose father was the first non-Indian to settle in the area, has dedicated 83 years to the accumulation of the pictorial and oral heritage of Cleveland and the surrounding region.

Likewise the Sapulpa Historical Museum has compiled the best archival collection of photographs and history available on the Red Fork and Glenn Pool region. Virginia Nevin Lane, the Museum's historian, and James W. Hubbard, a member of the organization's Board of Directors, were most cooperative in allowing the use of the photographs. In addition, they took their time to explain the background of each photograph.

Other individuals aiding in the research and collection of photographs for this work were: Leslie Brooks, Tulsa; Beryl D. Ford, Tulsa; Rex Hudson, Duncan; James O. Kemm, Tulsa; Dr. Guy Logsdon, Tulsa; P. C. Lauinger, Tulsa; Kelly Marshall, Ponca City; Malcoln Rosser, III, Duncan. Among other institutions contributing photographs were: History Room, Bartlesville Public Library; Leslie Brooks and Associates; Oklahoma Historical Society; Oklahoma-Kansas Division of the Mid-Continent Oil and Gas Association; University of Tulsa, McFarlin Library; and Western History Collections, University of Oklahoma. Corporations allowing the use of their photographs were: Continental Oil Company; Getty Oil Company; PennWell, and Halliburton Services.

Of major importance was the support received from Getty Oil Company and the Getty Refining and Marketing Company. Without their interest in preserving this significant aspect of Oklahoma's petroleum industry, this book would never have been completed. Special appreciation is extended to Jack D. Jones and Charles E. Smith of Getty Refining and Marketing Company for the tremendous amount of support they extended to this project.

Finally my thanks to my wife, Jayne, and my three sons, Douglas, Bryan, and Tyler for enduring another book with me.

CONTENTS

Preface and Acknowledgements iv
Introduction 1
I. Red Fork 3
II. The Sue A. Bland No. 1 10
III. The Red Fork Oil Field 17
IV. The Uncle Bill Lowery No. 1 27
V. Cleveland's Offspring 43
VI. The Ida Glenn No. 1 62
VII. Flush Production 73
VIII. Glenn Pool's Boom Towns 87
IX. Tulsa: Oil Boom Town 108
X. Tulsa: Oil Capital of the World 117
XI. Glenn Pool's Legacy 132

ILLUSTRATIONS

J. Paul Getty ... 5
Tulsa, 1889 ... 6
Oil spring near Tulsa ... 6
Spring pole drilling rig ... 7
Dr. Fred S. Clinton ... 8
John S. Wick, Jesse A. Heydrick, and others ... 11
Dr. Fred S. Clinton, Dr. J. C. W. Bland, and others ... 12
Sue A. Bland No. 1 blowing wild, Red Fork ... 13
Sue A. Bland No. 1 with pumping equipment, Red Fork ... 13
Dawes Commission, Muskogee ... 13
Cable tool rig, Red Fork ... 15
Homemade pumping unit in the Arkansas River ... 16
Oil well on sand bar in the Arkansas River ... 18
Robert Galbreath ... 19
Charles F. Colcord ... 19
Tulsa'a Frisco Railroad Depot ... 20
Cable tool drilling, Red Fork ... 20
Wooden storage tanks, Red Fork ... 21
Roustabouts at Red Fork ... 22
Dr. Fred S. Clinton's home in Tulsa ... 24
Tulsa's Eleventh Street Bridge ... 25
Uncle Bill No. 1, Cleveland ... 28
Kansas Torpedo Company shooter ... 29
Suspension bridge across Cedar Creek, Cleveland ... 29
Main Street, Cleveland ... 30
Cleveland during the oil boom ... 31
Moving a store to Cleveland by railroad ... 31
Cleveland Boiler Shop ... 32
Merchants Hotel in Cleveland after explosion ... 33
Addington Hotel, Cleveland ... 34
Odd Fellows Lodge, Cleveland ... 35
Whiskey Cave, Cleveland ... 36
Bootlegger in Cleveland ... 37
Cleveland Fire Department, 1906 ... 37
Church picnic near Cleveland ... 38
Roping contest at Cleveland, 1906 ... 38
Cleveland fire ... 39
Diamond State No. 4, Cleveland ... 39
Cleveland, 1908, after fire ... 40
Roofing a 55,000 barrel storage tank, Cleveland ... 40
Cable tool drilling rig, Cleveland ... 41
Maher-Miller No. 3, Cleveland ... 41
Cleveland's Main Street during oil boom ... 44
Dirigible near Cleveland ... 44
Cleveland gusher ... 45
Cable tool drilling rig, Cleveland ... 46
Mayfield No. 4, Cleveland ... 46
Wooden storage tanks, Cleveland ... 48
Bill and Sherman Copeland, teamsters, ca. 1904 ... 48
Constructing wooden storage tanks, Cleveland ... 49
Derrick destroyed by high winds, Cleveland ... 49
Sawmill, Cleveland ... 50
Pipeline crew, Cleveland ... 50
Boston Pool ... 52
Stout's Ferry, Cleveland ... 52
Wooden foot bridge over Arkansas River, Cleveland ... 53
Clayton Lucas operating Arkansas River ferry, 1924 ... 54
Osage Inn, Osage City ... 54
George F. Getty ... 55
Getty service station ... 56
Frank Buttram ... 57
Bridge crew, Blackburn pools ... 57
55,000 storage tank burning on A. H. Barnes farm ... 58
Toxway Oil Company earthen storage pit, Cleveland ... 58
Ida Glenn No. 1, Glenn Pool ... 63
Well being shot, Glenn Pool ... 63
Central Torpedo Company explosion, Glenn Pool ... 64
Shooting a well, Glenn Pool ... 65
Cable tool rig drilling, Glenn Pool ... 65
Drilling rig, Glenn Pool ... 66
Horse-powered pipe-pulling rig, Glenn Pool ... 67
Overview, Glenn Pool ... 68
Pumping equipment, Glenn Pool ... 68
Gypsy Oil and Gas Company wagon, Kiefer ... 69
Coal fired boiler, Glenn Pool ... 70
Gypsy Oil and Gas Company explosion, Glenn Pool, 1910 ... 71
Gusher, Glenn Pool ... 74
Storage tank construction, Glenn Pool ... 75

Line fight, Glenn Pool 76
Wooden storage tanks, Glenn Pool 77
Earthen storage pits, Glenn Pool 77
Polecat Creek on fire, Glenn Pool, 1908 78
Polecat Creek after fire, 1908 78
Glenn Pool fire 80
Preparing a fire dike, Glenn Pool 80
55,000 barrel tanks, Glenn Pool 82
St. Louis and San Francisco Railroad Depot, Kiefer 82
St. Louis and San Francisco Railroad loading rack, Kiefer 83
Sapulpa Refining Company, Sapulpa 84
Sapulpa Refining Company fire, 1911 85
St. Louis and San Francisco Railroad Depot, Sapulpa 88
Interior, St. Louis and San Francisco Depot, Sapulpa 89
Construction of interurban, Sapulpa 90
St. James Hotel, Sapulpa 90
Bovaird Supply Company, Sapulpa 91
Ben Hubbard, Bovaird Supply Company manager, Sapulpa 92
Dining Room, Harvey House Restaurant, Sapulpa 93
Dewey Avenue, Sapulpa 93
Whiskey still, Glenn Pool 94
St. Louis and San Francisco Railroad, Kiefer 95
Kiefer, 1906 95
Interior of general store, Kiefer 96
Charlie Shobe's and Bob Gaberdiel home, Kiefer 97
Main Street, Kiefer, at start of oil boom 98
Main Street, Kiefer, during the oil boom ... 99
Impromptu bank, Kiefer 99
Kiefer town band 100
Max Meyer, Amos Midman, Carl Morris (professional prize fighter), and Babe Banks 100
Petty and Sloat Livery Company, Kiefer ... 101
Kiefer fire, 1904 102
Main Street, Mounds 103
Elizabeth Winter's Boarding House, Taneha 103
Sherman Smith's gasoline service station, Taneha 105
Harry Young, Sabo Grocery delivery truck 106
Main Street, Tulsa, 1906 109
Main Street, Tulsa, 1907 110
Interurban car between Tulsa and Glenn Pool 110
Main Street, Tulsa, 1910 111
Cosden Refinery, West Tulsa 112
Prairie Oil and Gas Company's Tulsa warehouse crew 113
The Texas Company's West Tulsa Refinery 224
Skyline, Tulsa, 1939 115
First National Bank of Tulsa, ca. 1902 118
Railroad depot, Kiefer 119
Patrick C. Boyle 120
Waite Phillips 120
Philtower Building, Tulsa 121
Harry H. Rogers 121
P. C. Lauinger 122
Skelly Oil Company's Tulsa headquarters .. 123
Opening day ceremonies, IPE 123
W. G. Skelly addressing crowd at IPE 125
Night view, IPE, 1949 126
Petroleum Engineering Building, University of Tulsa 127
Robert M. McFarlin 129
McFarlin Memorial Auditorium, University of Tulsa 129
James A. Chapman 130
Indoor drilling rig, University of Tulsa 130
Tulsa Petroleum Club emblem 131
The Texas Company's pump station, Glenn Pool 134
Gulf Oil Company's pump station, Kiefer .. 134
D. W. Franchot Company's loading rack, Kiefer 136
Gypsy Oil Company's casinghead gasoline plant, Kiefer 136
Oklahoma Natural Gas Company plant, Sapulpa 138
Thomas Gilcrease 140
Alfred E. Aaronson 140
Thomas Gilcrease Institute of American History and Art, Tulsa 141

MAPS

Red Fork, Cleveland, and Glenn Pool, ca. 1907 . . . . Frontispiece
Oil and gas areas, Indian Territory, 1888 4
Red Fork Field, 1917 59
Pawnee County Oil and Gas Fields 60
Cleveland Field, 1917 61
Oklahoma pipelines 128
Glenn Pool, 1917 139

GRAPHS AND TABLES

Crude Production, Glenn Pool, 1907-1914 132
Oil production by Oklahoma, Texas, Arkansas,
Kansas, and Louisiana, 1906-1926 133
Natural gasoline production, Oklahoma
and Nation, 1909-1919 137
Capacity of Tulsa-Glenn Pool refineries, 1926 137-138
Marketed natural gas production, Oklahoma
and Nation, 1906-1915 141

THE RUSH BEGINS

INTRODUCTION

In 1901, the American petroleum industry was entering its fourth decade. In the previous 40 years most of the nation's production had been centered in the eastern part of the United States in the Appalachian and Lima-Indiana regions or on the West Coast of California. In the first year of the 20th century, however, the huge Spindletop Field was uncovered in Texas and oilmen began to turn their attention to the middle of the country. Within six months of A. F. Lucas' gusher at Spindletop, the Sue A. Bland No. 1 was completed at Red Fork in the Creek Nation. Its completion helped launch the beginning of Oklahoma's oil boom.

Like previous oil strikes in present-day Oklahoma, the Red Fork discovery was plagued by federal controls over drilling on Indian land and a dearth of readily available markets. While these drawbacks hindered continued development at Red Fork, the Sue A. Bland No. 1 turned the attention of many oilmen away from Texas and northward to Oklahoma and Indian territories. This was the field's greatest importance, because once oilmen became convinced that there was oil in the Twin Territories it became simply a matter of time before the true potential of the region's oil and gas deposits was tapped.

Stimulated by Red Fork, wildcatters scattered over both Oklahoma and Indian territories. Their persistance was rewarded when, in 1904, the Cleveland Pool was discovered in the northeast corner of Oklahoma Territory, about 30 miles northwest of Red Fork. Following close on the heels of Cleveland, the prolific Glenn Pool was located the following year, 1905, about 10 miles southeast of Red Fork.

With the three discoveries, Cleveland in Oklahoma Territory, and Red Fork and Glenn Pool in Indian Territory, Oklahoma's history was forever altered. Although there had been commercial discoveries of oil in Oklahoma previously, the string of finds starting with Red Fork and ending with Glenn Pool launched the oil boom era in the state, for their production attracted pipeline companies to the region. Oklahoma thus was linked to a vast network of pipelines offering a ready outlet for the state's oil. With an adequate means of marketing available, the Oklahoma oil industry boomed. Overnight, small rural communities were transformed to bustling oil boom towns in which millions of dollars changed hands daily. Fortunes were made, lost, and made again on the turn of a drill bit.

Glenn Pool thrust Oklahoma into the forefront of the American oil industry. In the three decades from 1906 to 1935, Oklahoma ranked first or second among the oil producing states for 25 years. The remaining five years it ranked third.

CHAPTER ONE

RED FORK

When J. Paul Getty, the founder of Getty Oil, recalled, in 1973 and 1974, his beginnings in petroleum industry, he credited his early years in Tulsa, Oklahoma, as giving him the experience necessary to succeed in such a highly competitive business. With the financial backing of his father, George F. Getty, the younger Getty entered the Oklahoma oil industry shortly after the mad scramble for riches was touched off by the discovery at Red Fork. One of thousands caught up in the rush, Getty quickly learned that the key to success was early and accurate information. Thus, as Getty explained, he haunted the halls and lobby of the Hotel Tulsa, the center of the petroleum industry in what then was known as the "Oil Capital of the World," "talking about oil, exchanging information, or 'dope' as we called it." "I learned very quickly," Getty continued "that the one who had the most information was the one likely to get the most information in return....If you never volunteered any, you never got any....So I would mill around and inquire about where persons had been and what they had seen....And in turn, they would ask the same questions of me." According to Getty, "it was an atmosphere in which a closed mouth did not pay off...the more you talked, the more you learned." "After my experiences in Tulsa," Getty declared "I knew all the nuts and bolts of the oil industry."[1]

The beginning of the boom which brought Getty to Tulsa started in 1901 at Red Fork, about 50 miles northwest of the Three Forks where the Arkansas, Grand, and Verdigris rivers join, in what was then the Creek Nation. Here—in an area bounded today by OK-91 Highway on the west, the Arkansas River on the north and east, and the northern boundary of Creek County on the south—the Oklahoma oil industry came of age. Although there had been previous oil discoveries in what became Oklahoma, none had attracted the nationwide attention that Red Fork generated. The result was the Oklahoma oil boom.[2]

Eventually the Red Fork District was expanded to cover all of T19N-R12E south of the bend of the Arkansas River, where it turns almost due west, as well as the E ½ of Section 25 and the NE ¼ of Section 36-T19N-R11E in Tulsa County. Although Sand Springs is located approximately in the center of the pool, Tulsa, about 10 miles to the east, is most commonly associated with the Red Fork strike. The name Tulsa was derived from the Creek village Tallahassee (Old Town), originally located along the banks of the Tallapossa River in the old Muskogean or Creek Nation, in present-day Alabama. Unfamiliar with the Creek dialect the early Spanish explorers misprounced the word as Tallasi.[3]

As early as 1826, a faction of the Upper Creek ceded portions of their homeland in exchange for a new home near the Three Forks area. At first the Lower Creeks refused to join the Upper Creeks. Only when rampaging whites burned many of their towns, including Tallahassee and another nearby village called Lochapoka, did they agree to migrate.[4]

A group of surviving Lower Creeks from Lochapoka arrived at the bend of the Arkansas River, near what became Tulsa, in 1836. Rebuilding their village, they named it Tallahassee. Like its predecessor in Alabama, the pronouncation of the new Tallahassee was corrupted, first into Tulsey Town and then Tulsa.[5]

The region gradually became a trading center for the surrounding countryside, and, after the the Civil War, Lochapoka Crossing or Tulsa became a major fording point on the Arkansas River for herds of Texas cattle being driven north to markets in Kansas and Missouri. During this boom period, Red Fork, four miles southwest of Tulsa on the south or opposite side of the Arkansas River, gradually developed into a sister settlement.[6]

Although Red Fork eventually boasted a small store and blacksmith shop, its growth stagnated in the late 1870s. In the 1880s, however, a new boom was created as railroads began to penetrate the area. On August 21, 1882, the Atlantic and Pacific Railroad, or A & P, reached Tulsey Town. Four years later, it had bridged the Arkansas River and extended its tracks to Red Fork and Sapulpa, which, for a while became the western terminus

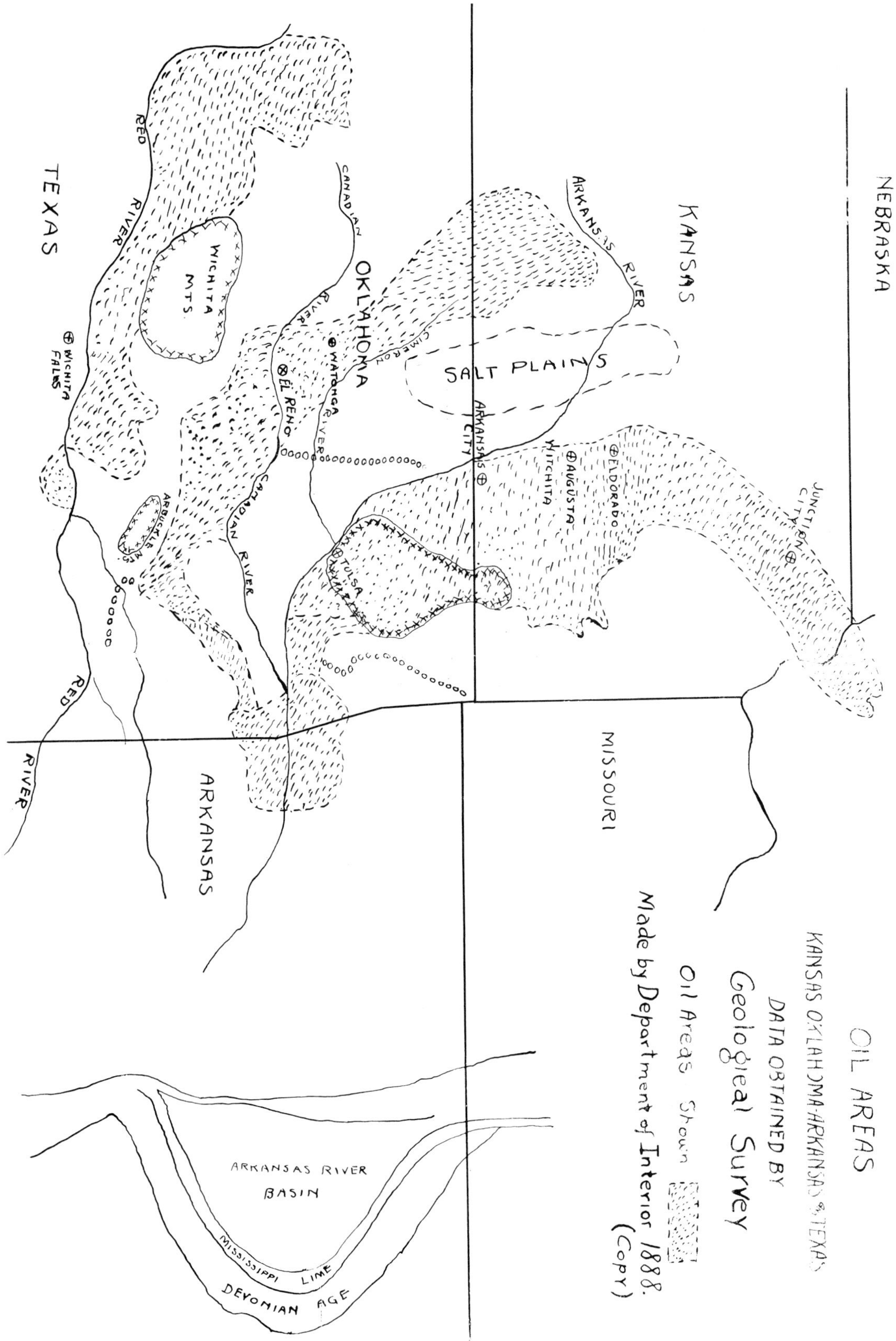

Released by the United States Department of the Interior in 1888, this geological map clearly shows the potential oil producing area centered around Tulsa. However, federal restrictions on the leasing of Indian land hampered the early development of the area. (Oklahoma Heritage Association).

J. Paul Getty, the founder of Getty Oil and pioneer Oklahoma oilman. According to Getty, he learned the "nuts and bolts of the oil industry" during his experiences in Tulsa at the beginning of the state's oil boom. (Getty Refining and Marketing Company)

of the line. Slightly more than a decade later, in 1897, the A & P line in Oklahoma was taken over by the St. Louis and San Francisco Railroad, or Frisco.[7]

It was the Frisco Railroad that officially gave the name Red Fork to the settlement. Originally, the railroad's charter specifically required the Frisco to cross the Arkansas River near what then was called the Red Fork of the Arkansas River, the present-day Cimarron River. However, the crossing at the Cimarron was not suitable. To solve the problem, Frisco officials ordered the bridge built several miles to the east at Lochapoka Crossing and then named the crossing Red Fork to comply with the charter requirements.[8]

With the arrival of the railroad at Red Fork, James H. Parkinson and J. M. Hall and Co., which later became H. C. Hall and Co., established two large general stores in the community. Wagon shops, three small hotels, two livery stables and feed stores, and a barber shop also were added, and a post office was established on January 3, 1884. Parkinson was the first Postmaster. When a local stockyard was built beside the Frisco Railroad tracks, Red Fork mushroomed into a major cattle shipping point, boasting direct connections with St. Louis, Missouri.[9]

At the same time that Red Fork was becoming a major cattle shipping center, America's oil industry was coming of age. As early as 1859, the same year that Colonel Edwin L. Drake completed his well at Titusville, Pennsylvania, Lewis Ross had accidently uncovered a shallow deposit of crude at the Grand Saline in the Cherokee Nation. Almost immediately numerous pioneer oilmen were attracted to Indian Territory. Among them were Robert M. Darden, who organized the Chickasaw Oil Company in 1872; H. W. Faucett, who examined portions of the Choctaw and Cherokee nations for potential oil deposits in the mid-1880s; and Edward Byrd, who drilled a well near Chelsea in the Cherokee Nation in 1886. That same year, another unnamed oilman drilled a 1,300-foot well about three miles west of Nickle Creek just to the south of Red Fork; however, he found no crude and abandoned the hole.[10]

It was not until 1894 that the Creek Nation began to draw the serious attention of oilmen. In that year, Michael Cudahy, a meat packer from Omaha, Nebraska, leased approximately 200,000 acres from Creek officials, and, after drilling two wells inside the city limits of Muskogee, moved his rig to the old well site on Nickle Creek and reopened the hole. Cudahy deepened the well to 1,750 feet before he also abandoned the site.[11]

Although most oilmen believed that the Creek Nation held potentially profitable petroleum deposits, Cudahy and the others were greatly handicapped in their efforts to tap the crude because of the question of who actually owned the land. The Creek Treaty of 1833 gave to the Creek Nation in fee simple what later became all of Creek, Okfuskee, Okmulgee, McIntosh, and Seminole counties; the southern two-thirds of Tulsa County, the western half of Muskogee County; the northern two-thirds of Hughes County; and all but the extreme eastern edge of Wagoner County. Later, present Seminole County, except for the extreme eastern part, was ceded to the Seminole Nation, but even so the Creek Nation controled a vast portion of east-central Oklahoma. Included within this grant of land was what became the Red Fork and Glenn Pool oil fields.[12]

However, the land was granted to "the Creek nation of Indians" and not individual tribal mem-

The Main Street of Tulsa, Creek Nation, in 1889. At the time the future "oil capital of the world" boasted a population of 200. (Beryl D. Ford).

An early-day oil spring near Tulsa. Such springs were valued for their curative powers. To provide easier access to the water at this site, stone steps were constructed and a pole placed at the bottom to provide a hand grip as cups were dipped into the water. (University of Tulsa, McFarlin Library).

A spring pole drilling rig. To construct such a rig a sapling near the drilling site was bent over and one end of the cable supporting the bit was attached to the end. The downward thrust was created by pulling the end of the sapling down and the upward pull was created by the natural straightening of the sapling to its original shape. Such simple rigs were used by early-day oilmen in Indian Territory prior to the 20th century. (Continential Oil Company).

bers. Prior to the allotment of individual plots, the Indians did not actually own the land but simply used whatever land they needed, with ownership remaining with the tribe as a whole. As a result, it was difficult to secure leases from individual tribal members. Likewise, whenever oilmen did secure leases from tribal authorities, the agreements often were vetoed by federal bureaucrats, who insisted that the United States government retained the right to control the Indians' property and that tribal treaties clearly excluded non-Indians from "the boundaries of the [Indian] nations."[13]

Many oilmen hoped that with the individual allotment of tribal lands under the provisions of the Dawes Act of 1887, interference from federal officials would cease. Unfortunately, even under the provisions of this legislation, fee simple title to the land was withheld from allottees for a pe-

As an ardent booster of Red Fork, Dr. Fred S. Clinton believed that the drilling of a successful oil well near the community would allow it to surpass its rival, Tulsa. (Oklahoma Historical Society).

riod of 25 years. Instead, the allotees were issued "patents" which allowed the "United States...[to] hold the land...in trust for the sole use and benefit of the Indian to whom such allotment shall have been made." The act went on to declare that if any conveyance "shall be made of the lands set apart and allotted...or any contract made touching the same, before the expiration of the time...such conveyance or contract shall be absolutely null and void." Thus, the Creeks were prevented from gaining absolute title to their land, and without an absolute title they could not lease their property. As a result, the legal status of oil leases in all of Indian Territory remained clouded.[14]

The hopes of oilmen rose with the passage of the Curtis Act on June 28, 1898, but again they were disappointed, for the Curtis Act effectively gave federal officials complete control over the continued development of the region's oil reserves. Under its provisions, the Secretary of the Interior was authorized "from time to time to provide rules and regulations in regard to the leasing of oil, coal, asphalt, and other minerals in...[Indian] Territory." The Curtis Act specifically provided that "such leases shall be made by the Secretary of the Interior, and any other leases for any such minerals othewise made shall be absolutely void." In addition, leases were restricted to 15 years and 640 acres.[15]

Moreover, according to the Curtis Act, individuals specifically were prohibited from entering into lease agreements. Under its provisions, it was "unlawful for any person,...except as hereinafter provided, to claim, demand, or receive for his own use or for the use of anyone else, any royalty on oil, coal, asphalt, or other mineral...any rents on any lands or property belonging to any one [tribal member]....'Also, all "royalties and rents" were "payable to the tribe."[16]

The rent was set for those holding "oil, coal, asphalt, or other mineral claim at the rate of one hundred dollars per annum, in advance, for the first and second years; two hundred dollars per annum, in advance for the third and fourth years, and five hundred dollars, in advance, for each succeeding year." Not only was this a large sum of money for the time, but also all the payments were held as credit on royalty due when the specified mineral deposit was developed and had to be paid whether the lease was developed or not. Failure to pay on schedule voided the lease and forfeited any money already paid. In addition, any damage incurred during the development of oil deposits within Indian Territory had to be paid for, with the value of the damage established by the Secretary of the Interior. The single concession to previously established oilmen was that any leases approved by Congress were allowed to continue in force for 15 years and "preference shall be given to such parties in renewals of such leases."[17]

It was not until March 1, 1901, when, by congressional action, the enforcement of the provisions of the Curtis Act relating to the leasing of mineral lands was eased. However, the reprive was short lived, for on June 30, 1902, Congress reversed itself. In a supplemental agreement with the Creeks, it amended the 1901 act to allow "Creek citizens...[to] rent their allotments [only], for strictly nonmineral purposes, for a term not to exceed one year for grazing purposes only and for a period not to exceed five years for agricultural purposes." The 1902 act went on to declare that "any agreement or lease of any kind or character violative of this...shall be absolutely void and not susceptible of ratification in any manner, and no rule of estoppel shall ever prevent the assertion of its invalidity."[18]

Despite the legal uncertainties involved in obtaining Creek leases prior to the admission of Oklahoma as a state in 1907, some pioneering oilmen were willing to gamble. The stakes were raised with the discovery of the rich Spindletop Field in Texas on January 10, 1901. Afterward, the focus of the oil industry was turned toward the Southwest. The question was where would be the most potentially productive place to drill.[19]

At the beginning of the 20th century, the geological analysis of potential drilling sites was still an almost unknown science. What geological knowledge that was applied was based on "creekology," a theory originally advocated by Mike Benedum, a famous pioneering wildcatter. Benedum's theory was derived from a combination of superstition, hunch, and geological knowledge gained from many years of firsthand experience. Basically creekology held that streams altered their courses to flow around an underground formation. With this in mind, Red Fork, which was centered in a huge bend in the Arkansas River, was an obvious choice.[20]

Adding to the theory of creekology was the more practical reason that the Frisco Railroad offered a means by which any crude located near Red Fork could be shipped to market. This was no small matter inasmuch as a great deal of the earlier petroleum development in the region was hampered by inadequate markets. Many of the finds that already had been located had not been profitable because their only outlet was local farmers, who used the crude for grease or as dip for their animals. Such a market easily was swamped by a single well.[21]

Another reason was the intense rivalry between Red Fork and Tulsa to be the economic center of the northwestern portion of the Creek Nation. Boosterism was rampant as the two communities, one south and one north of the Arkansas River, struggled to gain supremacy over the other. Caught up in the struggle were two of Red Fork's most outstanding citizens, Drs. John C. W. Bland and Fred S. Clinton. Bland's standing was heightened by the fact that his wife, Sue, was a citizen of the Creek Nation. In an aggressive attempt to bolster Red Fork's standing in the region, the two physicians decided to sink a well on the south side of the Arkansas River.[22]

As Clinton explained, "our decision to drill...was based on our faith in the development of resources of the Indian Territory, in industrial activity and oil interest in Kansas and Texas, the allotment of Indian lands, and the establishing of the legal right of ownership of property." Such factors, according to the doctor, "presented a golden opportunity for immediate progress if we could strike oil and secure national publicity without delay." Thus, according to Clinton, the two physicians decided to sink a wildcat well near Red Fork in the hope that any resulting discovery would make the community the dominant economic center of the area.[23]

Ironically, while the strike at Red Fork accomplished the desired results and eventually allowed Oklahoma to assume a position of leadership in the American petroleum industry, the hoped for boom passed Red Fork by after a few years. Instead, Tulsa, the community that Clinton and Bland set out to overshadow, reaped most of the benefits from the Red Fork discovery. While Red Fork languished, Tulsa grew into a major petroleum center, which boasted of being the "oil capital of the world." Within 26 years of the physicians' find, Red Fork disappeared. On October 11, 1927, Red Fork was annexed inside Tulsa's city limits and ceased to exist as an identifiable community.[24]

There are two accounts of the drilling of the Red Fork discovery well. One credits the two physicians, Bland and Clinton, with uncovering the field. The other story gives the credit of locating the pool to Jesse A. Heydrick and J. S. Wick, two experienced oilmen. Both claims have just arguments, and with the passage of time it is probably impossible to reach an objective conclusion as to the validity of either claim.[25]

CHAPTER TWO

THE SUE A. BLAND NO. 1

The controversy surrounding who drilled the first well at Red Fork began almost as soon as the discovery was made, with various historical accounts supporting the claim of either Jesse A. Heydrick and John S. Wick or Dr. Fred S. Clinton and Dr. J. C. W. Bland. To back Heydrick's and Wick's claim, the Heydrick family deposited an accumulation of documents in the Western History Collections at the University of Oklahoma. In response Dr. Fred S. Clinton countered by emphatically declaring in an article he wrote for *The Chronicles of Oklahoma,* the official publication of the Oklahoma Historical Society, that he joined with Dr. J. C. W. Bland to drill "the first oil and gas well on the Sue A. Bland homestead... adjoining Red Fork, Indian Territory." Because of the difficulty of determining just who drilled the first well, and inasmuch as all four men played a role in opening the field, both accounts are presented in this chapter.[1]

As early as November 9, 1899, the John S. Wick Company, ignoring federal regulations that virtually prohibited such an arrangement, entered into discussions with several prominent Creeks to lease tribal land for drilling purposes. Eventually, Thomas J. Adams, Wash Adams, Lewis Adams, James Sapulpa, William A. Sapulpa, Samuel C. David, John I. Yargee, L. C. Perryman, D. L. Berryhill, and Lilah D. Lindsey signed an agreement leasing Wick 410,000 acres. The property was bounded by on the Arkansas River on the north, the Frisco Railroad on the east and south, and the border of Oklahoma Territory on the west.[2]

At the same time, Wick entered into a partnership with Jesse A. Heydrick, a pioneering oilman from Butler, Pennsylvania. Heydrick, who was nearly 70 years old, had already spent 40 years in the nation's oil fields and brought a wealth of experience to the venture. In return for half interest in the John S. Wick Company, his task was to secure adequate financial backing for the project. Shortly afterward, although Wick and Heydrick had ignored federal regulations prohibiting their lease, they unsuccessfully attempted to have it validated in court.[3]

Unpreturbed with the apparent illegality of their act, Wick and Heydrick contracted with J. M. Dunn to examine their Creek lease and select the best possible drilling site. In the meantime, Heydrick had persuaded several Pennsylvanians to invest in the venture. Aiding Dunn in the survey was John Q. A. Kennedy, another Pennsylvania oilman, who had been sent to the area to represent the investors' interest in the operation.[4]

Although the Indians who had signed the lease were anxious for drilling operations to began, the undertaking was hampered by delays and Wick's apparent newly acquired concern over federal regulations. When the partners failed to commence drilling within six months of the signing, the original 1899 agreement was voided. However, on July 16, 1900, Wick negotiated a new arrangement signed by Sue A. Bland, Dr. John C. W. Bland, Timmie Fife, Eli E. Hardridge, J. H. Land, E. L. Simley, Albert P. Owen, and several other Creeks. To help finance the operation, 50 shares of stock, with a value of $100 each, were sold by Wick and Heydrick.[5]

With a new lease, Wick and Heydrick were ready to drill. As soon as Dunn and Kennedy had selected a site in the NW ¼ of the SE ¼ of Section 22-T19N-R12E as the most promising location, Wick and Heydrick contracted with the Crossman Brothers drilling company of Joplin, Missouri, to move their drilling rig from Sapulpa to Red Fork and begin operations. Situated on unalloted land on the northwest edge of the town of Red Fork, the site was between a quarter and a half of a mile from the Frisco Railroad tracks, approximately half a mile southwest of the present-day intersection of I-240 and US-75, just outside of the I-240 right-of-way.[6]

The Crossmans wasted little time in loading their equipment on a Frisco Railroad flatcar for transport to Red Fork. However, when the drilling rig arrived H. H. Adams, the local railway agent, refused to honor the Crossmans' New York bank draft as payment for the freight charges. At this point, according to the supporters of Wick and

John S. Wicks (third from the left, top row), Jesse A. Heydrick (on the left, front row), and various other Creeks who either signed or helped negotiate the lease for the John S. Wick Company to drill in the Red Fork area. Top row, left to right: John R. Yargee, Lewis Adams, Wicks, George W. Adams, James Sapulpa, and Timmie Fife; middle row left, to right: Legus C. Perryman, Joseph Henry Land, William A. Sapulpa, and Thomas J. Adams; front row, left to right: Heydrick and an unidentified Creek. (Oklahoma Historical Society).

Heydrick, the two oilmen enlisted the aid of Bland and Clinton.[7]

Just what the two physicians contributed is somewhat vague, but they apparently either agreed to cash the bank draft themselves, or, as Clinton claimed, borrowed $300 from the railway agent and then loaned the money to Perry Crossman, who actually handled the drilling for the brothers. Crossman later denied that he had borrowed any money from either physician. Nonetheless, the two oilmen apparently did receive enough money from Clinton and Bland to secure the relase of the drilling rig. In return, one account claims that Wick gave the two physicians one share each in the well, but this was not supported by the well's log, which lists only Heydrick, the Crossman Brothers, and Wick and Company as owners.[8]

Once the drilling equipment was released, Perry Crossman spudded in the well and began making hole on May 10, 1901. Within five days he had reached the 150-foot level. However, because it was being drilled on unallotted land, some of Heydrick's Pennsylvania investors became concerned over the lack of a clear title, and while the well was under way Heydrick was forced to return to Pennsylvania on May 15 to reassure his financiers. At the same time, Crossman left the drilling site for a short visit to Missouri. This left Wick in charge of the operation.[9]

Although two of the primary participants—Heyrick and Crossman—were absent, drilling continued with Luther Crossman, Perry Crossman's son, and Wick operating the rig. On June 24, 1901, shortly before midnight, the bit suddenly struck a "stray gritty lime" at the 534 foot level. Wick ordered Luther Crossman to continue

Dr. Fred S. Clinton (second from the right) and Dr. J. C. W. Bland (on the right) played a major role in the drilling of the Sue A. Bland No. 1, the discovery well at Red Fork named for Bland's wife. (Oklahoma Historical Society).

to make hole and when the bit pentrated the formation it released a pocket of natural gas and oil that shot 30 feet into the air.[10]

Wick was asleep near the rig when the well blew in. Awakened by the younger Crossman, he took one look at the gusher and then rushed to the nearest telegraph office. Although under strict orders to maintain secrecy in the event of a strike, Wick excitedly telegraphed Perry Crossman: "Send packer, Oil is spouting over the derrick." Crossman later accused Wick of being drunk when he sent the message, but the damage had been done. Telegraph operators from Red Fork to Joplin were jarred awake by the dramatic news—and the rush was on.[11]

Those supporters of Clinton and Bland tell a different story in which the two doctors were mainly responsible for the discovery. Both physicians were well-known area residents. John C. W. Bland had been born in Centerville, Iowa, in 1860, and after graduating from the Missouri Medical College in 1883, moved first to Dodge City, Kansas, and then the following year, 1884, to what became Tulsa, where he opened his medical practice. Eventually he married Sue A. Davis, the daughter of W. T. Davis, a prominent Creek. The couple had seven children: Vera, Era, Owen, Hazel, Arlies, Davis, and John. Involved in several ventures in the Red Fork-Tulsa area, Bland had been appointed postmaster at Red Fork, and in 1900, a year before the uncovering of the Red Fork Field, he had established a large general store in the community.[12]

Dr. Fred S. Clinton's father had first arrived in the Creek Nation in 1872 as a cowboy moving a herd along the Chisholm Trail from Texas. After the cattle drive, he took a job in a general store owned by F. B. Severs at Okmulgee, the capital of the Creek Nation. The elder Clinton eventually married Louise Atkins, a Creek citizen and local teacher, and founded the Half Circle S Ranch along Duck Creek, near Mounds on the present-day Creek-Okmulgee county line. In 1882 he moved his family to Red Fork, where his son, Fred, established his medical practice.[13]

Clinton contended that in May, 1901, Bland asked him to stop by his home one morning to discuss the potential of a 500,000-acre oil and gas lease, which according to rummor had recently been signed by several Creeks. While the prospect of a nearby oil discovery excited them, the two physicians realized that any such lease had to be approved by Secretary of the Interior Ethan Allen Hitchcock before it would be legal. During their conversation, Bland mentioned that the Crossmans' drilling rig was stranded at Red Fork's rail yard. As Bland explained, the freight agent would not allow the equipment to be unloaded until the bill was paid, and the Crossmans did not have enough cash to pay the bill.[14]

The Sue A. Bland No. 1, discovery well of the Red Fork Field, blowing wild. (Oklahoma Historical Society).

The Sue A. Bland after the derrick and drilling equipment had been removed and a pumping unit emplaced. (Oklahoma Publishing Company).

Members of the Dawes Commission meeting in Muskogee. Dr. Fred S. Clinton was able to persuade the Commission members to grant Sue A. Bland the allotment on which the discovery well at Red Fork was drilled. (Oklahoma Historical Society).

According to Clinton, it was he who suggested to Bland that they take advantage of the situation, see if they could acquire the services of the drilling rig, and start a well on some land that his wife, Sue, wanted for her allotment just to the east of Red Fork in the NW 1/4 of the SE 1/4 of Section 22-T19N-R12E. Clinton proposed that if Bland could get his wife to agree, he would raise the necessary money to pay the Crossmans' bill and see if the oilman would rent them his drilling rig. Realizing that if they struck oil it would give Red Fork "the widest publicity…attract oil people and insure development," Clinton readily agreed.[15]

As soon as Sue A. Bland acquiesced to the proposition, Clinton hurried to Adams, the Frisco Railroad's agent at Red Fork, and borrowed $300. After getting the money, Clinton approached Perry Crossman and persuaded the oilman to agree to the plan. Fortunately, according to Clinton, the $300 was enough to "free the equipment and get the drilling started." Hurriedly Crossman unloaded the rig, moved it to the selected site, spudded in, and began drilling on May 10, 1901.[16]

Drilling went slowly as Crossman encountered several unexpected problems. Of major concern was food for the drilling crew. Not only was food often difficult to acquire, but when it was available it was more expensive than anticipated. Good drinking water also was in short supply and expensive. In addition adequate housing for the drilling crew was hard to find. While some of the men were housed in Colonel Robinson's Hotel at Red Fork, others were forced to find rooms in private homes. If they were fortunate, the rooms also came with food. It was not uncommon for Crossman and some of the others to take their meals with Bland's married sister, Mrs. Vera Bland Stickles.[17]

Another problem was financing. To overcome the shortage of money, Bland paid the drilling crew with blocks of property he owned near the well site. It proved to be a profitable venture for the crew members as land values skyrocketed after the discovery. For example, in one instance two lots near the well site, which initially cost $1,400, were resold within 24 hours by the original purchaser, Frank O. Brown, for $2,800 each. Another invester offered Bland and Clinton $40,000 for the original well site.[18]

Although Bland paid the crew with nearby land, he needed additional cash. In his search for money, he offered several other prominent citizens of the area, including Dr. F. B. Fite, Creek Chief Pleasant Porter, and J. H. Hill, the opportunity to invest in the well. Unfortunately, because Sue A. Bland did not yet have title to the property, it was almost impossible to induce others to sink money into the project, and, in the end, Bland spent a considerable amount of his personal fortune to finance the project.[19]

In late June of 1901, Bland, who was overseeing the drilling operation, became ill with appendicitis. Nonetheless, he doggedly continued to maintain a watch over the work. However, as the completion of the well neared, the task became too strenuous for Bland and during the early morning of June 25, 1901, he sent for Clinton and asked him to "take full charge" while he rested. Rushing to the well site in response to Bland's request, Clinton was on hand shortly afterward when it blew in, shooting oil over the top of the derrick and soaking his clothes with crude.[20]

Once the well was completed the two divergent stories concerning the discovery of at Red Fork generally coincide. All four of the men— Wick, Heydrick, Bland, and Clinton— realized that they did not have a legal title to the property on which the well was located. However, Clinton was the first to act.[21]

With Bland incapacitated with appendicitis, Clinton obtained a power of attorney from Bland's wife Sue authorizing him to file "upon my allotment, for me and in my stead which has been selected by me….[in] the N.W. 1/4 of S.E. 1/4 Section 22, Township 19 N. Range 12 East," the quarter section in which the well was located. Immediately after having the power of attorney notorized, Clinton left for Muskogee, the headquarters for federal Indian officials in Indian Territory. With him he took a quart bottle filled with Red Fork crude.[22]

Riding his buggy across the Arkansas River to Tulsa, Clinton arrived at the railroad station just as the east bound train was leaving. Dropping his reins he dashed for the train and jumped on. He was in such a hurry that he was still wearing the oil-soaked clothes.[23].

Clinton arrived in Muskogee late in the afternoon of June 25, and rushed to the home of Dr. F. B. Fite, the personal physician of Allison Aylesworth, who, as a member of the Dawes Commission, was in the process of alloting individual Creeks' their share of tribal land. Elated to find

A typical cable tool drilling rig in operation at Red Fork in the winter of 1901-1902. To the right is the rig's walking beam, balanced on top of the sampson post. As the walking beam alternately was raised and lowered, it lifted and dropped the cable holding the drill bit in the hole. The pounding motion of the bit gradually pushed the hole deeper. (University of Tulsa, McFarlin Library).

Fite at home, Clinton hurriedly explained the reason for his sudden visit. However, Fite had never seen raw crude before, and he was unsure that the two physicians actually had struck oil. Realizing the importance of the discovery, Fite suggested that they test the contents of the quart jar to make sure that it was oil. Obviously the most practical way was to see if it would burn.[24]

Going outside to Fite's backyard the two men poured the contents of the jar onto some wood shavings. Fite then held a match to the wood, and stood back as it burst into flames. Returning inside the house, Fite and Clinton poured some of the remaining crude into a new lantern, took it back outside, and lit the wick. According to Clinton it "burned like good kerosene." After extinguishing the lantern, Clinton and Fite got into Fite's buggy and rushed to Aylesworth's home.[25]

On the way, Fite explained that although Aylesworth had been ill, he was feeling better and Fite was planning to release the Dawes Commissioner from his care the following morning and allow him to resume his duties. When Clinton and Fite arrived, Aylesworth listened to Clinton's story and then asked him to meet him at his office before 8:00 a.m. June 26, 1901. As Clinton later recalled "we were there promptly," and Aylesworth had already made "all proper preparations for filing my power of attorney for the Sue A. Bland's application for homestead allotment."[26]

As soon as he had filed Sue A. Bland's request for allotment with Aylesworth, Clinton started back to Red Fork. Just prior to leaving he stopped at the office of Fite and Dr. J. L. Blakemore to thank Fite for his help. Clinton did not personally know Blakemore; however, Blakemore was F. B. Severs' physician, and Clinton's father had worked for Severs when he had first arrived in Okmulgee in 1872. When the two men realized that they had a mutual friend, Blakemore asked permission to join Clinton on the return trip to Red Fork. Clinton agreed and Blakemore suggested that instead of taking the train that they rent a hack. To have someone to help drive the buggy, the two physi-

After the opening of the Red Fork Field, several local citizens hurried to sink additional wells. Many wells were primitive affairs, such as this homemade pumping unit operating in the Arkansas River near Tulsa. (Sapulpa Historical Museum).

cians invited A. Z. English, Severs' son-in-law, to join them.[27]

The three men left for Red Fork about 4:00 p.m. on June 26 on what was known as the Trail Road. Unfortunately, they continued their trip after sundown and in the darkness somehow got off the trail and entangled in a barbed wire fence near the IX Ranch. One of their horses was badly cut. However, the ranch was owned by Bluford Miller, Severs' brother-in-law. Miller gladly replaced the injured horse, and, by travelling all night, Clinton, Blakemore, and English reached Red Fork early the morning of June 27.[28]

Through his efforts, Clinton had acquired legal title to the land on which the Red Fork discovery was made. The actual legal description of Sue A. Bland's allotment was lots number 1 and 6 in block number 4; lots 1, 2, 3, 5, 6, and 7 of block 27; lots 1, 2, 3, and 5 of block 28; the north half of lot 3 of block 30; and lot 1 of block 31 inside the townsite of Red Fork. The remainder of her allotment included the NW 1/4, SE 1/4, less 5.93 acres of the St. Louis and San Francisco Railroad in Section 22-T19N-R12E. Elated with the news that Aylesworth had approved of his wife's allotment, Bland named the well the Sue A. Bland No. 1 in her honor.[29]

CHAPTER THREE

THE RED FORK OIL FIELD

When Clinton returned to Red Fork with Sue A. Bland's application for allotment firmly filed with federal authorities in Muskogee, he found "a seething mass of people over every portion of ground or space that could be occupied." During the time that both physicians had been away—Bland was ill and Clinton was in Muskogee—many "different people who felt at liberty assumed authority to run the business [the well]." Fortunately, Bland quickly recovered and resumed command of the operation.[1]

Many local citizens scrambled to see the well. Several of the early arrivals were relatives of Clinton and Bland, including Bland's wife; his daughters, Vera and Era; and his son, Owen. Clinton's mother, Louise, his brothers and sister, Paul, Lee, and Vera, and Paul's wife, all visited the Sue A. Bland No. 1. Their interest was more than just passing, for several of them claimed allotments just to the east of the discovery well. Other spectators included: H. H. Adams and his wife; Lincoln Postoak; J. B. Hall; J. W. Hall; Frank O. Brown and his wife; Mr. and Mrs. H. E. Bridges; C. M. Forsythe; R. D. Atkins and his wife; Mr. and Mrs. H. D. Lindsey; Robert and Thomas Hughes; Mrs. Robert Fry; May and Lena Sanger; Rena and Sammy Norman; and Colonel Robinson, Edward Robinson, and George Robinson. Among the more interested visitors were Mr. and Mrs. John I. Yargee and Van and Pleas Yargee, who owned property just to the west of the Sue A. Bland.[2]

Because of Wick's indiscretion in telegraphing Crossman of the strike, word of the discovery spread rapidly. Fred S. Barde, the local correspondent for the *Kansas City Times,* hurriedly dispatched an account of the discovery to his newspaper. "OIL WELL GUSHER FIFTEEN FEET HIGH" the headline proclaimed. Other journalist were quick to pick up the story. "A GEYSER OF OIL SPOUTS AT RED FORK" headlined the *Tulsa Democrat.* The *Muskogee Weekly Phoenix* declared that the Sue A. Bland No. 1 indicated a "strata of oil-bearing sand [that] extends under thousands of acres." Continuing, the Muskogee newspaper, obviously much to Clinton's and Bland's disgust, proclaimed Tulsa as the center of the new field.[3]

The discovery could not have occurred at a more opportune time, for the oil industry was still excited from the Spindletop discovery and there was no lack of potential investors. Rummors that proclaimed Red Fork to be "mightier than Spindletop" or that "Red Fork would surpass the great wild Beaumont as a center of action and easy money" simply fanned the flames. Within a short time thousands were hurrying to the strike[4]

As early as the morning of June 26 the roads between Tulsa and Red Fork were jammed with people. As one journalist recorded, "a never-ending stream of wildly excited men are flocking [to]...the scene of the great strike." Although Red Fork was only two and a half miles southwest of Tulsa, the only means of getting from one town to the other was across the railroad bridge over the Arkansas River, and it could carry only pedestrian traffic. Those with wagons and teams had to cross on a ferry operated by Lee Clinton, Fred's brother. Those with sufficient money simply rode the Frisco train to Red Fork and then walked to "see the well." "Half of Tulsa was standing around gaping at the spouting oil well," one witness recorded.[5]

Others came from throughout the Twin Territories. To accommodate the increase in traffic, a special railroad car had to be added to the train leaving Oklahoma City bound for Tulsa. Among those Oklahoma Citians "prepared to make investments" at Red Fork were Charles F. Colcord, Robert Galbreath, Usher Thomas, Boston Wilson, Lee Van Winkle, Taz Upshaw, John Carson, and William Pettee. Among those who followed Clinton back to Red Fork from Muskogee were Drs. Blakemore and Fite and J. H. Hill. Even Pleasant Porter, Chief of the Creeks, made the trip.[6]

As Clinton later recalled, within a short time the community's "few impoverished buildings, barns and horse lots were overrun by a mob of sightseers, promoters, speculators, gamblers, bad girls, doctors, lawyers and Indian chiefs." Would be oilmen poured not only into Red Fork but also

An oil well on a sand bar in the Arkansas River near Tulsa. It was not uncommon to drill directly in the river bed; however, it could be dangerous as heavy rains far upstream often caused the river to flood without warning. In this instance, a tree trunk has been hauled onto the sand bar to provide footing for the pumping unit. (American Petroleum Institute).

into the nearby communities of Tulsa and Sapulpa. By the afternoon of the day following the completion of the well, all livery rigs in both communities had been rented by oil scouts, and late comers were forced to walk to the well site.[7]

Red Fork was converted almost overnight from a small farming community, with several stores and maybe a score of houses, into a bustling oil boom town. So many people crowded into the community that "a restaurant was erected immediately at the well to feed the people," and the town quickly took on a "carnival" atmostphere. Tents sprouted everywhere, and by July 4, 1901, more than 1,000 representatives of various oil firms were operating in the area. Several were local enterprises, such as the Red Fork based Superior Oil and Gas Company, formed shortly after the strike by Charley Donahue, Charley Collier, Ed Boneker, and Jim Mackin.[8]

Unfortunately, many of these ventures were purely speculative and others were completely fraudulent. "The woods are full of people trying to get leases," one witness recorded. Attorneys descending upon the town, lured by the scent of "future fees in the legal difficulties [that are] sure to arise," found a willing clientele, for many investors were unsure "over the legal status of [local] property rights" and wanted clear title to the land before they would agree to risk their money.[9]

Within 48 hours of the completion of the well an estimated 4,800 people were milling around Red Fork. In addition, "Telegrams are pouring in from all over the country" inquiring about the discovery. Heydrick himself was somewhat taken aback when he returned from Pennsylvania by train on June 29. So many people already had poured into Red Fork that he was forced to take a room in Sapulpa's Gladstone Hotel because there was no other space available. "Everything went," one local witness recalled, "the sky being the limit,

Robert Galbreath, who later discovered Glenn Pool, was one of the early arrivals in Red Fork after the completion of the Sue A. Bland No. 1. (Sapulpa Historical Museum).

Charles F. Colcord, an Oklahoma City businessman, hurried to the strike at Red Fork, formed a partnership with Robert A. Galbreath, and drilled several wells in the field. (Oklahoma Heritage Association).

for the boom was on and the restless humans hot in pursuit of the elusive goddess of chance were fairly tumbling over one another in their undignified but eager scramble for anticipated wealth." Not all vistitors were impressed with the scramble for riches, however, and one described the community as "one of the vilest spots in the [Indian] Territory."[10]

The hoard of prospective oilmen descending on the Creek Nation caused some concern among federal officials. Clifton R. Breckenridge, a member of the Dawes Commission, hurriedly informed Secretary of the Interior Ethan A. Hitchcock, of the strike. Wasting little time, Hitchcock telegraphed J. George Wright, the United States Inspector of Indian Territory, and ordered him to "investigate the report promptly." Wright contacted Guy P. Cobb, the inspector who was closest to Red Fork at the time of the discovery, and dispatched him to the well site.[11]

Cobb arrived at the Sue A. Bland No. 1 about 3:00 p.m. on June 27—just three days after the well had been completed. Reporting to Wright that there had indeed been a strike, he recounted that he had witnessed "a considerable flow of gas issuing from the top of the casing pipe." The escaping natural gas was accompanied "by a bubbling, rumbling, sound, and at intervals of perhaps from 15 to 30 minutes each, a small quantity of oil was forced up by the gas."Although previous newspaper articles had claimed that the oil was gushing over the top of the derrick, Cobb informed Wright that the crude only pushed "a distance of perhaps two feet above the top of the pipe" and then a part of the oil fell "back in [the hole] and part...[slopped] over the sides of the pipe."[12]

Tribal officials also began to take a dim view of the strike. Although Chief Porter previously had visited the well site, he later declared that "I think it unfortunate that oil, in apparent commercial quantities, has been found." Apparently Porter's main concern was that the discovery would "embarrass the work of appraising the allotments." While Porter acknowledged that "those who live in the supposed oil field and who have taken their allotments and have received certificates" have an undisputed title to the land, he

Tulsa's Frisco Railroad depot was the jumping off place for many oilmen rushing to Red Fork. From the depot, they rode wagons, such as these, to the railroad bridge across the Arkansas River, walked over the bridge, and boarded another wagon for the field. (Beryl D. Ford).

A typical cable tool drilling rig in operation during the boom at Red Fork. The rough cut timber used to build the derrick usually was cut nearby and untrimed logs were used for foundation supports. (Sapulpa Historical Museum).

Much of Red Fork's early production was stored in wooden tanks such as these. Because it was difficult to construct large wooden tanks that would not leak, each well usually was served by several smaller tanks called a battery. In this photograph the drilling equipment has been replaced by a pumping unit; however, the wooden derrick has been left in place. (Sapulpa Historical Society).

feared that "the Untimely Location of the Valuable Product, Before the Titles to the Allotments are Clear," will undoubtedly cause problems, and he wanted to insure that the claims of various Creek citizens to the property would not be ignored.[13]

Despite the boom, Heydrick was somewhat disappointed when he arrived back at the well site from Pennsylvania on June 29. "When let loose...[the Sue A. Bland No. 1] flows high into the air but soon blows out," he wrote to his son James A. Heydrick. The elder Heydrick was convinced that the driller had not located a rich producing sand. Instead, he believed that the well was capable of only producing about 10 barrels per day. Accordingly, when he received an offer of $20,000 for 40 acres of land near the well site, Heydrick recommended to his Eastern financial backers that the property be sold. However, they refused.[14]

To encourage potential investors, and thereby provide additional funds for continued drilling, Clinton, Bland, Wick, and Heydrick devised a method to keep up the public interest. It bordered on outright deceit. By placing a canvas bag over the well's bailer, and then lowering the contraption to the bottom of the well they were able to hold the escaping oil and gas in check. After enough pressure had built up to throw the crude high into the air, the bailer and cavnas bag were hurriedly drawn up the hole. With the blockage removed, the crude, pushed by the gas pressure, rushed up the pipe and was thrown as high as 10 feet into the air.[15]

As a result of such theatrics, several oilmen

began additional wells in the Red Fork area in spite of the question of unresolved legal titles. Their efforts received a boost when the federal government sold the townsite lots of Red Fork at auction in the spring of 1901. One of the first developers was Robert A. Galbreath, who later would uncover the rich Glenn Pool oil field.[16]

Galbreath had first visited the region in the company of several Oklahoma City businessmen who had hurried to Red Fork as soon as they had learned of the strike. In conjunction with Colcord and C. G. Jones, Galbreath purchased between 25 and 30 of the Red Fork lots at the government's sale. Colcord, Galbreath, and Jones also were successful in obtaining leases on all the land owned by the John Yargee family.[17]

Within a short time, Colcord, Galbreath, and Jones had moved two drilling rigs to the Red Fork region. One well was spudded in on a farm owned by Mrs. John Yargee and the other on the Van Yargee allotment. The three men started drilling without federal approval, and, as Galbreath later recalled, "pretty soon along came the Indian police, departmental sleuths, carpetbag scouts and uniformed bands of brass collars, forbidding and warning against further drilling.... It looked like the penitentiary for us," he continued.[18]

Fortunately for the oilmen, Chief Porter intervened in the matter. Although Porter was concerned about the development of the region by oilmen, Yargee's wife was his sister, and Porter used his influence to have Colcord's leases approved. Because of Porter's influence, Colcord, Galbreath, and Jones were allowed to continue through what Galbreath described as a "twelve-months occupancy feature of the homestead regulations which seemed to give Yargee the right to lease." As a result Colcord and the others had the jump on other Red Fork oilmen.[19]

Once Porter's influence had ended any chance of federal interference, Galbreath had hurried to Oklahoma City, purchased a train car load of used pipe for the Yargee well from Andy Binns, and quickly shipped the pipe to Red Fork. Within four months of the completion of the discovery well, another 10-barrel-per-day producer was completed. Encouraged, and protected somewhat from Indian officials because of Porter's intervention, the three oilmen drilled several other wells on the Yargee property. One proved to be a 40-barrel-per-day well from 1,200 feet. In February, 1904,

Two typical roustabouts at the time of the Red Fork discovery. The cable tool rig was so named because the drill bit was suspended in the hole by a large cable, shown here in the background wound around the rig's bull wheel. (Sapulpa Historical Museum).

Colcord, Galbreath, and Jones completed their No. 1 Nathaniel Yargee, which flowed 125 barrels every 24 hours. It not only extended Red Fork's productive area to the northeast of the community, but also attracted the attention of Jimmie O'Neal, President of the Prairie Oil and Gas Company, who traveled to Red Fork to see the strike for himself.[20]

Many of the early wells at Red Fork were gassers. At the time it often was difficult to sell the gas production. To open new markets for their natural gas, Colcord and the others made arrangements to pipe the gas to Tulsa where it could be used for lighting and heating. However, just before the operation started, Colcord, Galbreath, and Jones sold their Red Fork interests to Glenn

Branden, who in turn used the supply of natural gas to begin what became the Oklahoma Natural Gas Company. The three oilmen also developed a more profitable outlet for their natural gas wells. After locating a nearby deposit of building stone, which conveniently could be converted to cement by burning, they quickly built a cement plant and used the natural gas to reduce the stone.[21]

While Colcord, Galbreath, and Jones were able to utilize their influence with Chief Porter to pave the way for their drilling operations, other area oilmen were not so fortunate. The Dawes Commission was convinced that the many Creeks filing for allotments in the Red Fork area in June, 1901, did so at the urging of "speculators, who desired to secure leases from the citizens who might secure...[allotments] in that vicinity." "All possible care" must be exercised, the commissioners declared, to insure "that the best interests of the Creek citizens were subserved."[22]

Inspector Wright also warned that "any leases for the purpose of extracting...oil would not be recognized by the department [of the Interior]." Thus, as early as June 19, Wright was calling Chief Porter's attention to sections of the Creek Treaty of March 1, 1901, which discussed the question of individual allotments. The inspector pointed out that while Section 6 confirmed all allotments made "prior to the ratification of this agreement as to which there is no contest, and which do not include public property," Section 7 clearly specified that "Lands alloted to [Creek] citizens...shall not in any manner whatsoever, or at any time, be encumbered, taken, or sold to secure or satisfy any debt or obligation contracted or incurred prior to the date of the deed to the allottee."[23]

Section 7 went on to restrict the alienation of any allotments "by the allottee or his heirs at any time before the expiration of five years from the ratification of [the agreement]...except with the approval of the Secretary of the Interior." In addition, 40 of the 160 acres of each individual allotment were to be selected as the allottee's homestead, and this property was to be "inalienable and free from any incumbrance whatever for twenty-one years." Section 37 of the 1901 agreement also specifically addressed the question of leases and included several restrictions on the leasing of land. Among them was a requirement that leases were limited to "a term not exceeding one year, and after receiving title thereto without restriction."[24]

Chief Porter agreed with Wright that the terms of the 1901 agreement were specific and declared that "It appears to me that these sections are plain enough for a person of ordinary intelligence to understand, and will require no elucidation." The chief continued, notwithstanding his intervention on the behalf of Colcord, Galbreath, and Jones, to contend that "each and every one" of the leases acquired by "various methods by...persons with a view to securing the control of the future disposition of selected lands of the Creek Nation...are in violation of the terms" of the Creek Treaty of 1901. Porter went on to point out that he wholeheartedly supported the restrictions prohibiting the alienation of land by Creek citizens for a specified number of years. The allottees must be "taught the value of their lands" before they could function outside of tribal laws, and such a prohibition was needed to keep the Creeks from being cheated by non-Indians, he maintained. It was a policy with which Secretary of the Interior Hitchcock agreed.[25]

Such an attitude by tribal and federal officials continued to cause potential investors to hesitate before spending money at Red Fork. Eventually, the concern became so great that Heydrick was forced to suspend his drilling operations because of a lack of adequate funding. The fears were magnified in early August, 1901, when Wright made the opinions public knowledge.[26]

At that time, he clearly stated that "the lands of the [Creek] tribe...shall be allotted to the citizens of such tribe so as to give each an equal share of the whole in value." Wright went on to point out that while all allotments by the Dawes Commission prior to the ratification of the 1901 agreement were binding, "lands allotted...shall not in any manner whatsoever...be encumbered, taken or sold...except with the approval of the secretary of the interior." Those who compiled the 1901 Creek Agreement, he maintained, obviously wanted "to prevent the Creeks from disposing of the mineral in or under their lands unless they receive a reasonable compensation."[27]

Secretary Hitchcock, according to Wright, had decided that it was his prerogative to "determine whether or not such compensation is reasonable and just." As a result, "in the case of a valuable oil well...being discovered in or under any of the

Dr. Fred S. Clinton's home in Tulsa. Like many other Red Fork oilmen, Clinton eventually moved to Tulsa. Ironically, Tulsa was the city that he originally had hoped to overshadow with his well at Red Fork. (Oklahoma Historical Society).

Creek lands prior to the final appraisement of such lands, the value of such mineral deposit will be taken cognizance of by the...[Dawes] Commission...in appraising the lands." Should the commissioners decide that "the land upon which such deposit exists be valued at more than the standard value of an allotment, the [Creek] nation would have a lien upon the rents and profits...until the excess had been paid." In other words the royalty would accrue to the entire tribe, until the value of the land equaled the value of all other allotments, which was pegged at $6.50 per acre for 160 acres by the 1901 agreement.[28]

To clarify the matter of oil leases once and for all, Wright pointed out that Secretary Hitchcock had ruled that "a Creek citizen who attempts to make a lease by the provisions of which he disposes of...oil or other mineral, in or under his allotment, illegally attempts to dispose of a part of the fee." Wright took the opportunity to publicly declare that the Department of the Interior "will not, at this time, permit any citizen of the Creek Nation, or other person, to take or extract from the lands of the Creek Nation, in any manner whatsoever...oil." With such a pronouncement, many prospective investors had second thoughts about leasing Creek lands which they might loose at the whim of federal bureaucrats.[29]

While Bland and Clinton were safe in their operation on the Sue A. Bland lease, the 410,000 acres of land leased by John S. Wick Company were seriously in doubt. Obviously the few members of the tribe that signed the agreement could not claim anywhere near that amount of land, even if everyone of them took their 160 acre allotment within the bounds of the original lease. The company's chances of keeping its holdings were diminished on August 13, 1901, when Inspector Wright, declared that in compliance with federal law, the Department of the Interior was required to "put each [Creek] citizen in unrestricted possession of his land, and remove therefrom all per-

The Eleventh Street Bridge that connected Tulsa with the strike at Red Fork. Originally constructed as a toll bridge, the structure helped launch Tulsa toward becoming the "oil capital of the world." (Oklahoma Historical Society).

sons objectionable to him." In accordance with that law, Wright gave notice that "all persons in possession of lands who have not leases with the citizens who have filed on or selected such land, must remove therefrom not later than October 1, 1901."[30]

Under Wright's proclamation, anyone wishing to continue drilling operations at Red Fork after October 1, 1901, had to "make satisfactory leases with the citizen [holding the allotment to the land]." In spite of an appeal to Secretary Hitchcock on their behalf, Wick and Heydrick had their lease voided. Fearing the same result, many oilmen decided to bypass Red Fork. It was not until 1902 that the Secretary of the Interior approved the first oil leases in the Red Fork area.[31]

Although Bland and Clinton held title to the Sue A. Bland lease, it did not prove to be extremely profitable. Clinton maintained that his efforts in the project were motivated more by a desire "to find some self-supporting industry" that could be developed in Red Fork rather than becoming an oilman. "I did not expect to share his [Bland's] property," Clinton declared. "I never claimed or received one dime from the production of the well or wells drilled.... It was understood that I did not even want to be reimbursed for money or any aid furnished," he continued.[32]

Likewise, Bland did not reap a tremendous profit from his discovery. At the time of the completion of the Sue A. Bland No. 1, the nearest refinery was at Neodesha on the Verdigris River in Wilson County, Kansas, approximately 95 miles northwest of Red Fork. While the refinery offered $1.00 per barrel for the Sue A. Bland No. 1's crude, it cost Bland 90¢ a barrel to ship the oil. This left a 10¢ per barrel profit. The initial daily production of the discovery well was approximately 100 barrels, so Bland could expect a profit of not more than $10 daily, less other expenses of operating the well.[33]

Nonetheless, on July 5, 1904, Sue A. Bland was able to secure a lease with the Indianola Oil and Gas Company, which was subsequently approved by the Secretary of the Interior for a period of 15 years. Six years later, on February 18, 1910, she deeded her allotment to her heirs—Vera, Era, Owen W., Hazel M., Arlies S., Davis M., and John C. Bland. The next day, February 19, 1910, Sue A. Bland died. When the 1904 lease expired on July 4, 1919, it was renewed "for a term as long as oil or gas or either of them, is produced in paying quantities from said land."[34]

Because of the lack of a suitable profit from the crude, and the difficulty in gaining clear title to the property, the Red Fork Field never became a prolific producer and within half a decade was overshadowed by the tremendous production of Glenn Pool. Although not in the same category as some of Oklahoma's great oil fields, Red Fork's annual output, in 1950, was a respectable 15,330 barrels of crude. Of more importance than production, however, was the fact that the strike did generate enough interest to maintain the search for crude that eventually thrust Oklahoma, and Tulsa, to the forefront of the American petroleum industry.[35]

One of the most important outgrowths of the Red Fork discovery was the linking of Tulsa and Red Fork by a means other than the railroad bridge across the Arkansas River. Soon after the completion of the Sue A. Bland No. 1, Tulsa's Commercial Club began planning for another structure that would accommodate wheeled traffic as well as pedestrians between the town and the oil field. Most of the expenditure was underwritten by Don Hagler, George Williamson, and Melvin Laird, three leading Tulsa businessmen. Three years after the Red Fork discovery, in 1904, the Eleventh Street Bridge was dedicated. The linking of Tulsa with the oil fields south of the Arkansas River played a major role in the transformation of the community into the "Oil Capital of the World."[36]

CHAPTER FOUR

THE UNCLE BILL NO. 1

While the question of the leasing of Indian land at Red Fork was being decided by federal and tribal officials, oilmen turned their attention westward to Oklahoma Territory. In 1889 the federal government had embarked on a policy of opening the region to non-Indian settlement. Thus, through a series of land runs, lotteries, sealed bids, and tribal allotments the Unassigned Lands; the Cherokee Outlet; the Cheyenne and Arapaho; Comanche, Kiowa and Apache; Wichita and Caddo; Kaw; Tonkawa; Pawnee; Ponca and Oto-Missouri; Iowa; Sac and Fox; Kickapoo; Pottawatomie and Shawnee; and the Osage reservations, as well as the Big Pasture were removed from the public domain. With the passage of the Organic Act this region was joined with No Man's Land on May 2, 1890, to form Oklahoma Territory. Six years later, on March 16, 1896, the United States Supreme Court attached old Greer County to the new territory.[1]

The Organic Act eliminated the confusing question of land leasing that had prevented the development of Indian Territory's oil reserves. Under its provisions, homesteaders received a clear land patent to their claims. They had complete freedom in disposing of their property through lease or sale.[2]

With the attention of the petroleum industry focused on the the strike at Red Fork, it was only natural that oilmen would begin to examine neighboring Oklahoma Territory. Their attention was drawn to the Watt Armstrong farm, about three miles north of Granite in Greer County, where in 1901 the Oklahoma Natural Gas, Light and Heat Company completed an 8 to 10 barrels-per-day well. Although, some promoters proclaimed the discovery a "rival [of] the famous Spindletop gushers in Texas," the well was a primitive affair.[3]

To get the crude to the surface, the driller attached a rope to a sheet-iron bucket which was then lowered into the well, filled with crude, and then pulled to the surface by a horse and pulley. Nonetheless, A. J. Greiner, Secretary-Treasurer of the Oklahoma Natural Gas, Light and Heat Company, boasted that "Oklahoma [Territory] is bound to be one of the greatest oil fields in the United States." Local enthusiasm was so great that a refinery was built at nearby Gotebo to handle the well's output. Unfortunately, the strike's production did not live up to expectations and no subsequent development took place.[4]

Another attempt was made to drill on the east side of Cache Creek near Fort Sill. However, local homesteaders opposed to the drilling burned several derricks and the effort was abandoned. Another attempt was made near what locally was known as Asphaltus Spring, located approximately four miles southeast of Fort Sill. Here several wells were drilled to a depth of about 600 feet. Although "frequent layers of oil sand" were encountered, the activity was stopped because of law suits brought by area homesteaders.[5]

Afterward, most oilmen shifted their activity to northeastern Oklahoma Territory. As early as 1902 one well about nine miles southeast of Newkirk was reported to be producing "small quantities of gas and a high-grade crude oil." Its success encouraged the drilling of several other wells in the general area, but not enough oil was located to touch off a boom. Additional exploration took place near Guthrie, Oklahoma City, McCloud, Shawnee, Chandler, and Cushing, but again no large pools were uncovered.[6]

Oklahoma Territory's petroleum pioneers eventually began to concentrate their activity along the Arkansas River, just across the border from the northwest corner of the Creek Nation. Here the wide bends of the Arkansas River, which formed the northern and eastern boundaries of Pawnee County, offered the same general geological environment as the Red Fork area. But, equally important, Pawnee County had the advantage of being under the control of Oklahoma territorial officials and not subject to the rules and regulations of the Department of the Interior.[7]

Most of the county was covered with rolling hills, interspaced with long stretches of sand along the bed of the Arkansas River. In keeping with the prevailing geological theory of creekology, oil-

In the upper right of this photograph is the Uncle Bill No. 1, discovery well of the Cleveland Field. This photograph was taken after the completion of the well. Note the school children watching the burning oil in Cedar Creek, which separated Cleveland from the well site. (Robert Jordan).

men naturally were drawn to the sandy bends in the Arkansas River. In particular, they were attracted to the town of Cleveland, originally named Herbert in honor of the townsite owner, Willis H. Herbert.[8]

Initially, the county had been given to the Pawnee tribe as a reservation when they had fled Nebraska after almost being exterminated by the Sioux. Although the Pawnees were split over the question of removal, the first members of the tribe reached Indian Territory in the autumn of 1873, and, because they were distantly related to the Wichitas, were temporarly settled on the Wichita Reservation. The following year, on October 8, 1874, nearly 1,500 Pawnees who had remained in Nebraska joined the earlier arrivals.[9]

Shortly afterward, the tribal leaders selected approximately 282,020 acres of land between the Arkansas and Cimarron rivers as their permanent home. Unaccustomed to the new climate, many of the Pawnees sickened and died in their new land. From the 2,376 tribal members alive in 1873, only 804 remained in 1890, and on October 31, 1891, only 820 Pawnees were available to accept their individual allotments under the provisions of the Dawes Act. Afterward, the remainder of the Pawnee Reservation was sold to the federal government for $1.25 per acre.[10]

The excess land was opened to homesteaders at the same time the Cherokee Outlet and Tonkawa Reservation were opened by land run on September 16, 1893. With the abolishment of the reservation, the area was incorporated into Q County, Oklahoma Territory, with Pawnee named as the county seat. Shortly thereafter, on April 14, 1894, the name of Herbert was changed to Cleveland in honor of President Grover Cleveland.[11]

Many of the early residents were aware of po-

A shooter for the Kansas Torpedo Company loading nitroglycerin from an isolated storage shed. Note the special wagon used for transporting the nitroglycerin. It has separated padded compartments to prevent the explosive from being shaken. Such a shooter shot the Uncle Bill No. 1 on July 23, 1904, and brought in a gusher. (Oklahoma Historical Society).

The suspension bridge across Cedar Creek that connected Cleveland with the wells south of town. Note that many of the trees and much of the undergrowth along the banks of the creek have been killed, either from the crude that leaked from nearby wells or from burning oil on the surface of the water. (Robert Jordan).

tential oil deposits. According to Robert O. Jordan, the son of the first non-Indian to settle in the area, oil and salt water seeps were well-known along Cedar Creek south of Cleveland. Later, during the oil boom, Cedar Creek often was covered with burning crude set on fire by farmers wanting to keep the oil away from their land.[12]

Thus, it was not suprising when, in 1904, the Minnetonka Oil and Gas Company secured leases along Cedar Creek. After examining the area, the company decided to drill along the creek bank on the farm of William Lowery, almost in the center of the large bend of the Arkansas River. Because Lowery was better known as "Uncle Bill," the well was dubbed the Uncle Bill No. 1.[13]

Nearly 100 of Cleveland's 1,000 citizens were on hand when W. J. Fellows and John Schell spudded in the Uncle Bill No. 1 on May 27, 1904. However, the local Board of Trustees for the community did not share the same enthusiasm as many of the local residents. On June 10, 1904, less than two weeks after the Uncle Bill No. 1 was started and more than a month before it was completed, Cleveland's Board of Trustees enacted an ordinance that prohibited the drilling of oil or gas wells and the construction of oil storage facilities within the city limits.[14]

Because it had been started before passage of the ordinance, work on the Uncle Bill No. 1 continued. With 2-men crews working 12-hour shifts, the hole was deepened rapidly, and within a short time it was reported that the well had penetrated a formation that was "identical with that of developed fields." At the 600-foot level, sand filled with salt water and a trace of oil was located. Encouraged, Fellows and Schell hurried the work.[15]

On June 28, 1904, a month and a day after the well was started, a "light flow of gas" was found between 900 and 1,000 feet. The next day, at a depth of 1,100 feet, a third pay sand was found,

The Main Street of Cleveland prior to the oil boom. Because Cleveland was a well-established community before the discovery of oil, the town did not experience as much oil field violence as some other boom towns. Note the combination furniture store and undertaker parlor on the left. (Robert Jordan).

As soon as word of the Cleveland strike spread, the town underwent a dramatic building boom. Under construction in this photograph is one of Cleveland's typical oil field hotels. Nothing but a wooden shell, covered with sheet iron and separated into individual cubicles, such hotels offered little other than basic shelter. Notice that the front porch is supported by rough cedar logs. (Triangle Heritage Association of the Cherokee Strip).

Store buildings were in such short supply during the Cleveland oil boom that structures were shipped to the town from other nearby communities by railroad, unloaded, and opened for business. (Robert Jordan).

The Cleveland Boiler Shop. Such businesses were vital to the surrounding oil fields. Note the completed boilers on the right and those still under construction on the left. With no sidewalks, entrance to the building was across the two boards leading from the doorway to the street. (Robert Jordan).

and "with it came gas in abundance." The flow was strong enough to spew "ten or fifteen feet above" the well head.[16]

Wanting oil, not gas, Fellows and Schell continued to push the hole deeper. By the time the bit had penetrated another 150 feet—for a total depth of 1,250 feet—it was estimated that the Uncle Bill No. 1 was flowing between "three-quarters of a million to three million cubic feet" of natural gas daily. Such production created quite a stir among area oilmen, and when word of the find leaked out, representatives of several companies were dispatched to Cleveland to keep watch on developments.[17]

A fourth, stronger, gas zone was located on July 2, and the well's daily estimated natural gas production jumped to between 10 million to 20 million cubic feet. By this time, some oilmen began to believe that Fellows and Schell had located a large field of natural gas, and that the Uncle Bill would never produce oil. However, as the tools were being lowered into the hole to resume drilling after the well had been bailed, someone noticed that a film of oil covered the cable. Shortly thereafter, Mrs. E. G. Todd suddenly was sprayed with crude as she was holding her hand over the well head to feel the escaping natural gas. As the crew hurriedly pulled the tools from the hole, the Uncle Bill No. 1 came in as a 10-barrel-per-day producer.[18]

Although oil had been found, Fellows and Schell decided to drill deeper in the hope of increasing the well's output. By July 21, 1904, the Uncle Bill No. 1 had reached the 1,625-foot level, with the bottom 55 feet containing two rich oil formations, which were named the Cleveland and Kelso sands. The Cleveland Sand proved to be almost 20 feet thick.[19]

Because many of the early wells were plagued with parafin, which clogged both the tubing and the bottom of a well, and because crooked holes sometimes caused early drillers to miss rich pools by a few feet, it was a common practice to "torpedo" a well to increase its output. In theory the explosion, with the force of the blast directed downward and outward, not only would clear out deposits of paraffin but also would fracture the oil-bearing formation and stimulate production.

Because of the hasty construction, accidents were inevitable. This is the remains of the Merchants Hotel in Cleveland after natural gas, which had accumulated in the structure, was ignited on January 20, 1916. Natural gas was a constant danger in Cleveland inasmuch as many drillers simply vented it into the air because there was no ready market. Sometimes the gas would concentrate in low-lying areas and then explode with no warning after being ignited by a stray spark. (Triangle Heritage Association of the Cherokee Strip).

Those who practiced this dangerous and specialized practice were called "shooters."[20]

When nitroglycerin was substituted for gunpowder in 1869, the shooter's task became more dangerous. Afterward a typical shot utilized between four to six quarts of nitroglycerine compounded with various amounts of gunpowder or guncotten. The spilling of a few drops sometimes touched off a premature explosion with fatal results.[21]

With the idea of increasing the Uncle Bill No. 1's output, Schell contacted J. C. Moore, a shooter with the Kansas Torpedo Company, to shoot the well on July 23, 1904. Moore first filled the metal torpedoes with the explosives, and then lowered them into the hole. Anchoring them at the desired depth, he attached the triggering device, called a "go devil," to the anchoring cable and then dropped the go devil down the hole.[22]

According to one witness, when the torpedo and go devil struck one another, the explosive charge detonated with a muffled roar and "the shot tore the casing...and blew part of the packer in the bottom clean to the top." Production from the Uncle Bill No. 1 immediately jumped to nearly 250 barrels of crude daily. Unfortunately, it quickly fell to between 100 and 150 barrels for a short time, and then decreased once again before stabilizing at about 50 barrels of crude per day.[23]

Although the strike was not the magnitude of Spindletop, it did generate quite a bit of excitement among oilmen. Several local residents quickly formed the Cleveland Oil, Gas, and Manufacturing Company and began leasing nearby property. At first, wells were drilled in all directions from the Uncle Bill No. 1, but after several dry holes the oilmen discovered that the field lay to the south and west of the original strike. Sections 16 and 17

Cleveland's Addington Hotel contained one of the finest restaurants in Oklahoma Territory and maintained a strict dress code for its customers. Note the expensive automobiles parked in front of the establishment. (Robert Jordan).

of T21N-R8E proved to be the richest acreage in the pool. Other producers were completed in sections 18, 20, and 29.[24]

The effect of the Uncle Bill No. 1 was predictable and, according to Territorial Governor Thompson B. Ferguson, "felt all over the territory." Cleveland rapidly grew from a hamlet of less than 1,000 people to a booming oil town of more than 7,000. By early September of 1904, deposits at local banks had increased 75 percent. Within six months it was reported that 25 new homes and several new business buildings were under construction. It was estimated that a total of $1,500,000 was spent developing the Cleveland Field—quite a respectable sum for the early 20th century.[25]

With the oilmen came an assortment of merchants and legitimate businessmen, as well as gamblers and prostitutes, hoping to secure their share of the wealth. A land boom, fueled by "speculators, 'law-shysters,' prospectors, and 'fly-by-nighters,' " soon was under way and the price of leases soared. In spite of the building boom, new arrivals in Cleveland reported that "Hotels are badly overcrowded and it is almost impossible to secure accommodations." Many of the early arrivals were lodged in the Arlington Hotel.[26]

The structure originally had been built in St. Louis, Missouri, as a part of the Louisiana Purchase Exposition and World's Fair of 1903-1904. When the exposition closed, the building was purchased by Dr. G. W. Sutton, one of Cleveland's early-day citizens. Sutton had the structure dismantled and shipped to Cleveland on the Missouri, Kansas, and Texas Railroad. When the building arrived it was reassembled and opened as the Arlington Hotel.[27]

The Arlington boasted of one of the finest restaurants in the territory. Strict rules were enforced by the maitre d'hotel. No one was admitted unless properly dressed in a coat. However, such a rule was difficult to enforce in an oil boom town, and the restaurant kept a supply of coats near the front door so customers could borrow one if necessary. Meals were expensive, costing as much as $2.00 each, a high price for the time.[28]

Cleveland's population swelled even more in August and September of every year when the community hosted a series of horse races. Held at a race track to the southwest of the town, the races attracted a considerable following from Tulsa and other nearby towns in Indian Territory. Not only could visitors enjoy the races in Cleveland, but the town was in Oklahoma Territory, where, prior to statehood, liquor was legal as opposed to

The Odd Fellows Lodge on Main Street in Cleveland was constructed in 1888. By the time of the oil boom, part of the structure had been taken over by an oil lease broker. Note the barrels stacked in the alley against the building. They were used to store the homemade whiskey sold in the town's saloons. (Robert Jordan).

Indian Territory, which was "dry." During his early career, Will Rogers performed during the races.[29]

Many individuals took advantage of Cleveland's location to smuggle illegal whiskey into the Creek, Cherokee, Osage, and Pawnee Indian lands, and the community long had been a major supply point for bootleggers. With the hoard of oil workers came an even greater boom in the liquor traffic. According to Jordan, at the height of the oil boom the community boasted 13 saloons and 2 distilleries inside the city limits and several others outside.[30]

Most of the whiskey was manufactured locally. The standard recipe, according to Robert L. Lumsford, III, the third generation of his family to be engaged in the banking business in Cleveland, included chewing tobacco or creosote, to give the mixture the correct color; sulfuric acid, to give the drink a "kick;" and 120 proof alcohol. The ingredients were mixed together in a 120 gallon barrel and allowed to age while waiting behind the bar to be served. It was a common sight along Cleveland's streets to see large stacks of 120-gallon barrels waiting to be filled.[31]

After 1907 and prohibition, many of the bootleggers moved into the natural caves along the Arkansas River. Although statehood hindered the traffic in illegal whiskey in Cleveland, during the oil boom there was no shortage of liquor in the town. One of the most famous sources of illicit liquor was a partially completed building between Cedar Creek and the railroad tracks south of the town. Bootleggers would hide their merchandise in the old structure, and those wanting a drink would wander by several times a day to make their purchases.[32]

Most of the two-story saloons were concentrated in a two block area along Cleveland's main street. The establishments also doubled as brothels. While whiskey was served on the ground floor, the prostitutes were housed upstairs. To provide patrons entrance to the brothels without being

Whiskey Cave in the Osage Reservation across the Arkansas River from Cleveland. Cleveland was in "wet" Oklahoma Terrtitory, but federal law prohibited the sale of liquor across the river in the Osage Reservation. The prohibition was impossible to enforce as bootleggers made legal purchases of whiskey in Cleveland and then smuggled it across the river onto Indian land. They were aided by numerous caves, such as this, which offered ready hiding places for the illegal liquor. (Robert Jordan).

seen, stairs were built in the alleys behind the saloons.[33]

Because Cleveland was a well-established community at the beginning of the oil boom, it did not suffer as much violence as did some other oil boom towns. In one instance, however, a saloonkeeper, who suspected another local man of sharing the affections of his girl friend, caused quite an uproar. Once his suspicions had been aroused, the saloonkeeper watched the other man closely. One day the saloonkeeper, in the hope of catching his girl friend and rival together, took his pistol and rushed up the back stairs of the saloon to the girl's room.[34]

As he burst through the door, the saloonkeeper shot at his rival, who was nude, but missed. Fleeing from the room, without waiting to put on his clothes, the terrified man fell down the back stairs, ran through the side of an out-door privy, fled down the town's streets, and ran into the surrounding woods near the Arkansas River. Two woodcutters later reported that they had seen a naked man running through the trees toward the Osage Nation.[35]

During the height of drilling activity, according to Jordan there were between 150 and 200 prostitutes plying their trade in Cleveland. While city officials did little to hinder their activity, they did insist that the prostitutes keep off the streets. In fact, they only were allowed to leave the brothels daily between 2:00 and 4:00 p.m. This, city officials reasoned, would allow them sufficient time to do their shopping; however, to insure that proper decorum was maintained, the prostitutes were required to wear ankle-length dresses whenever they appeared outside.[36]

A short distance west of Cleveland a small community named Dog Center was established. The settlement acquired its name from the large number of dogs kept by the oil field workers, who made up most of the town's citizens. A typical oil field boom town of clapboard and canvas structures housing establishments catering to the drilling crews, Dog Center did not survive the boom. Eventually most of the community's residents either moved to another oil strike or to Cleveland. Some of the town's businesses actually were loaded onto railroad cars, hauled to Cleveland, and reopened.[37]

By July of 1905, barely a year after the original Cleveland discovery, Governor Ferguson reported 220 producing oil wells at Cleveland, with another 7 wells flowing gas. At that time only 28 holes had proved to be dusters. Many of the early wells were drilled close together on town lots, apparently inside the city limits in spite of the previously passed ordinance prohibiting such activity. Total production had reached almost 11,000 barrels of oil and approximately 11 million cubic feet of natural gas daily.[38]

The pool eventually was expanded to cover approximately 50 square miles in eastern Pawnee County. By 1914, Cleveland area oilmen had tapped the deeper producing horizons in the field: the Cleveland Sand, found at a depth of between 1,500 and 1,700; the Bartlesville Sand between 2,400 and 2,500 feet; and the Tucker Sand be-

After statehood, Cleveland's saloons were forced to close. Nevertheless, an illegal whiskey trade flourished. One of the well-known spots frequented by bootleggers was a partially completed building south of the main part of Cleveland between the railroad tracks and Cedar Creek. The building was filled with hay and rubble, thus offering numerous hiding places for whiskey bottles. Men would stop by, pay for their whiskey, and then the bootlegger would retire inside the structure and bring out their liquor. In this photograph the bootlegger is doing a booming business. (Robert Jordan).

The Cleveland Fire Department, 1906. The man holding the reins is George Lanning, Cleveland's town marshal and the person charged with maintaining law and order during the early oil boom. (Robert Jordan).

Cleveland was fortunate in that much of the vice associated with some other oil boom towns bypassed the community. Because Cleveland had been established prior to the oil boom, churches and other civilizing effects led the crusade to keep the worst vice and violence under control. This is a picnic organized by Cleveland's Methodist, Baptist, Christian, and Presbyterian Sunday Schools. (Triangle Heritage Association of the Cherokee Strip).

Because liquor was legal in Cleveland, the town became a favorite recreation area for many residents of nearby Indian Territory. Of special attraction was the local racetrack, where spectators could gamble and drink without interference. This photograph is of a roping contest held at the Cleveland race track in November, 1906. (Robert Jordan).

Although many of the buildings in the older sections of Cleveland were constructed of native stone or brick, most of the structures built during the oil boom were wooden. After the wood dried they became fire hazards. Here three men, who have moved their chairs off the front porch of the Dunlop Hotel, watch part of the town burn. (Robert Jordan).

The Diamond State No. 4, to the southwest of Cleveland, blowing in as a gusher. Note the corrugated steel oil storage tank to the right of the well. Most of the early storage tanks either were wood or bolted iron. (Robert Jordan).

tween 2,600 and 2,800 feet. In addition, the Skinner Sand, a prolific producer of natural gas, was located at the 2,100- to 2,200-foot level.[39]

By the spring of 1917, several oil companies were involved in the development of the productive acreage in T21N-R8E. A 250-barrel-a-day well was brought in by the Schoenfelt Oil Company in the SW ¼ of Section 17, and the Scioto Oil Company drilled a 100-barrel-per-day producer in the Arkansas River bed, just to the north of the town of Cleveland in Section 5. Most of the wells had an output less than that of the Scioto well; however, the Franchot Oil Company brought in two wells in Section 17 with an output of 30 to 40 barrels daily. Likewise the Oklahoma Oil Company completed a 50 barrel well in Section 35, and the Piqua Oil Company and the State Tads Oil Company also had some small producers in the pool.[40]

Several other nearby pools were located in the scramble to drill after the original discovery. The South Cleveland or Olney Pool was found about six miles to the southwest of the town of Cleveland in 1915. The discovery well, drilled by Charles Page in Section 34 T21N-R8E, flowed 2 million cubic feet of natural gas daily from 1,400 feet. Two pays of oil also were found—Red Fork Sand and Bartlesville Sand. Productive acreage was quickly developed to the north and east of the discovery well in sections 25, 26, 27, and 35, and to the south and west in sections 2, 3, 4, 5, 8, 9, 10, and 11 of T21N-R8E. Eventually the South Cleveland Pool was incorporated into the Cleveland Field.[41]

Many of Cleveland's early wells "produced an immense volume of gas." For example, the No. 3 Diem, in Section 18 flowed 25 million cubic feet daily. Unfortunately, while driving through the Cleveland Field in January, 1913, an official of the United State Bureau of Mines reported that "many

Downtown Cleveland in 1908 after a fire had gutted the local post office. Note that the rest of the block was saved because the post office was next to a building built of native stone, which isolated the blaze. (Robert Jordan).

As production from Cleveland's wells grew, the need for larger storage facilities became acute. Here a roofing crew is sealing the top of a 55,000-barrel steel storage tank near Cleveland. (Triangle Heritage Association of the Cherokee Strip).

A typical cable tool rig operating in the Cleveland Field. Note the "black dogs" hanging from the derrick. They were kerosene lamps used to light the rig so that drilling operations could continue throughout the night. Apparently this well has been completed and the crew is preparing to set the casing pipe stacked near the derrick. (Triangle Heritage Association of the Cherokee Strip).

When the Maher-Miller No. 3 blew in as a gusher, it attracted a number of onlookers, including several women from Cleveland. Note the wire fence surrounding the drilling site. Because Cleveland was the center of a cattle raising area, it was necessary to fence many drilling sites to keep out wandering animals. (Robert Jordan).

of the wells were heard blowing oil and gas into large flow tanks." "The gas," he explained "could be seen shimmering from the top of the flues on the tanks and often it was so rich in the heavy vapors it would sink to the ground." He estimated that the amount of natural gas wasted by venting it into the atmosphere approached 20 million cubic feet daily by January, 1913.[42]

The problem at Cleveland was no ready market for the pool's crude production, and no market whatsoever for any natural gas output. With no outlet, the gas was worthless to oilmen. As a result, the natural gas simply was vented into the air.[43]

Because much of the natural gas was heavy, "the valleys or gullies through some of the richer parts of the field were full of...gas." It often hung "like a fog over the ground." Such a condition could be very dangerous for travelers and there were several instances reported of automobiles or open flames igniting pockets of natural gas, which exploded without warning.[44]

Also the lack of adequate shipping facilities forced many Cleveland oilmen to store their production in open pit earthen tanks. This, of couse, allowed large amounts of oil to be lost through seepage or evaporation. The problem was eased somewhat in late 1904 when the the Prairie Oil and Gas Company, a subsidiary of Standard Oil of New Jersey, connected the field to its pipeline and shipping facilities at Red Fork. At the same time nearly 1,500,000 barrels of steel and wooden tankage were being built in the Cleveland area. As a result of this construction, the price of oil properties jumped.[45]

Because Oklahoma Territory oilmen were not hindered by federal bureaucrats and tribal officials, Cleveland was the first fully developed pool in what became Oklahoma. With its growth came an influx of capital necessary to continue the development of the region's oil and gas reserves. Likewise, the discovery greatly stimulated the search for additional crude—a search that launched Oklahoma on the road to becoming a leader in the nation's oil and natural gas industries.[46]

CHAPTER FIVE

CLEVELAND'S OFFSPRING

Cleveland's valuable production quickly led to additional drilling away from the initial discovery. The first offshoot of Cleveland was the Boston Field, located just across the Arkansas River in Osage County. It was opened in 1904 by the Gypsy Oil Company and actually was a northwest extension of the Cleveland Pool. Although small in area, the Boston strike was prolific, with some wells flowing 6,000 barrels of oil and 10 million cubic feet of natural gas daily. Because of the timing of its discovery, the same year as the opening of the Cleveland Field, and its closeness, the Boston Pool greatly stimulated the Cleveland oil boom.[1]

Three ferries across the Arkansas River, two located just to the east of Cleveland and the other north of the town, joined the community with the Boston find. Stout's Ferry, to the east of Cleveland near a natural ford across the river, was the most prosperous of the three. It was operated by Tom Peck.[2]

Just across the Arkansas River, about a mile east of Cleveland, the community of Osage City was established in 1906. It was connected to Cleveland by the Missouri, Kansas, and Texas Railroad, and, along with Cleveland, served as the major supply point for the Boston discovery. Osage City's most famous establishment was the Osage Inn, an exclusive restaurant with an outstanding reputation throughout the area.[3]

The Cleveland discovery also spawned several other finds in Pawnee County. Shortly after the completion of the Uncle Bill Lowery No. 1 in 1904, the Price Pool was located about 10 miles to the south of Cleveland in sections 25 and 36 of T20N-R7E. However, it proved to be insignificant find, and the area was abandoned in 1948.[4]

The next strike was made just to the west of the town of Ralston, in sections 2, 3, and 4 of T23N-R5E. The discovery well, a wildcat drilled by J. M. Critchlow and his partners in 1910, was located in Section 3. Later the New England Oil Company reported a 50-barrel-a-day well to the west, in Section 4. Five horizons eventually were found, with some of the pool's 11 producing wells flowing 190 barrels daily.[5]

In 1912, just to the north of the strike at Price, the Terlton Pool was located in sections 19 and 30 of T20N-R8E, near the town of Terlton and along House Creek. Two other nearby strikes were made in 1917: the Terlton North Field, located in the four corners area of townships 21 and 22 North and ranges 7 and 8 East, and the Terlton Southeast Pool in sections 33 and 34 of T20N-R8E. Some wells in the area were huge gassers, with one Selby Oil & Gas Company well flowing 5 million cubic feet of gas every 24 hours.[6]

In 1914 the Quay Field was discovered in Section 4-T20N-R7E of Pawnee County. However, it later was expanded southward into Payne County and of its 307 wells, only 57 were spread along the southern border of Ranges 5 and 6 East in Pawnee County. Even so it ranked second in production only to Cleveland among Pawnee County's fields.[7]

There were two auxiliary pools to the original discovery at Quay. On February 22, 1922, the Marland No. 1 School Land well, in the NW ¼, NW ¼, SE ¼ of Section 33-T20N-R8E, located the West Quay Field. One well, an offset to the discovery well, was completed as a 398 barrel-per-day producer. Nearly three decades later, in 1955, the Quay North was opened in Section 11-T21N-R5E. Initial production from the Quay North was 5,623 barrels annually from two wells.[8]

The Watchorn or Morrison Field was located by Frank Buttram, who was working for the Fortuna Oil Company at that time, in sections 3 and 4 of T22N-R3E, just to the southeast of the town of Watchorn in extreme northwestern Pawnee County. The discovery well, the George L. Miller No. 1, was completed on December 27, 1915, as a 35 million cubic-feet-per-day gasser. During the field's first seven years of life, production was limited to natural gas; however, on July 14, 1922, the George L. Miller No. 3 was completed as an oil well. When deepened in October of 1923, it flowed 650 barrels of oil daily. Production in the Watchorn Field peaked in 1925 at 2,232,080 barrels annually.[9]

Four satellite fields of the original Watchorn

Teamsters on Cleveland's Main Street during the oil boom. Cleveland became the major supply center for a number of other fields located in Pawnee County. (Robert Jordan).

As the center of oil development in the region, Cleveland attracted a number of entrepreneurs catering to the oil field workers. Here the owner of a dirigible demonstrates his craft to a crowd of onlookers. (Triangle Heritage Association of the Cherokee Strip).

strike, all in T22N-R3E, were developed. Watchorn East was opened in 1942 to the northeast of the original pool, along the southern edge of the Arkansas River. Two others were opened in 1945, the Watchorn Northwest in sections 19 and 20 and the Watchorn North, directly north of the 1915 discovery. The Watchorn West was located in 1953 to the west of the Watchorn Field, along the Pawnee-Noble county line. Of the four satellite Watchorn Pools, the Watchorn East proved to be the most productive.[10]

The Lauderdale and Ranch Creek pools also were found in Pawnee County in 1915. The former was located by Charles Page southeast of the Cleveland Field in a long bend of the Arkansas River running through the southeast corner of T21N-R8E, the northeast quarter of T20N-R8E, and the west central portion of T20N-R9E. Drilled in Section 34-T20N-R8E, the discovery well was completed at 1,400 feet as a 2 million cubic-feet-per-day gasser. Lauderdale eventually became one of the most prolific pools in Pawnee County. The second pool, located in 1915, was the Ranch Creek Pool. Located by the Uncle Sam Oil Company in Section 20-T21N-R7E, it proved to be a minor discovery.[11]

Although several small oil wells and two gas wells had already been drilled in the area, in 1916 the Republic Oil & Pipe Line Company located the Jennings Field with the completion of a 200 barrel-per-day producer in Section 21-T20N-R7E. Two nearby wells drilled in Section 25, one by the Carter Brothers Oil Company and the other by Frank Brown, proved to be small producers. Maximum reported oil and natural gas production was 400 barrels and 3 million cubic feet daily. Even so, additional drilling expanded the pool into sections 15, 16, 21, 22, 23, 26, 33, and 34 of T20N-R7E, and across the county line into Creek County. Most of the production was located in Creek County, with only 11 of the field's 95 productive wells found in Pawnee County. In 1917, the year following the discovery at Jennings, another small field, the Casey Pool was located in Section 21-T21N-R6E.[12]

Just to the north of the Jennings Field, the Turner Investment Company located the Hallett Pool. The discovery well, drilled on the McMillan farm in the NE ¼, SW ¼ of Section 3-T20N-R7E, was completed on April 6, 1922, as a 60- to 75-barrel-per-day-producer. As the field expanded

A gusher shooting oil into the air near Cleveland. This well had just been shot with nitroglycerin to increase its production. (Triangle Heritage Association of the Cherokee Strip).

over much of the west-central part of T20N-R7E, many of the wells blended into the northern extremity of the Jennings Field. As a result many oilmen considered the two fields as a single pool. An extension, Hallett Northwest, was located in Section 30-T21N-R7E in 1949, but it proved to be a small find.[13]

After the initial strike at Cleveland, oilmen drilled extensively to the north of the strike. The Duquesne Petroleum Co., for example, completed the Chas. Colclazer No. 1 in the SW ¼, NE ¼, SW ¼ of Section 8-T21N-R6E on August 21, 1923, with an initial flow of 660 barrels daily. However, production quickly dropped off and continued drilling to the Wilcox Sand produced no additional oil. As a result, it was not until 1951 that the nearby Casey North Field was uncovered

A typical cable tool drilling rig in operation near Cleveland. Note the pile of coal in the left foreground. It was used to fuel the boiler located in the shed just behind the pile. (Triangle Heritage Association of the Cherokee Strip).

The Mayfield No. 4 in the Cleveland Field. It appears that the rig builders have just completed work on the derrick and are posing for the photographer on the cross beams and crown block. (Triangle Heritage Association of the Cherokee Strip).

a mile to the south in Section 17-T21N-R6E. It too proved to be a relatively small producer.[14]

The Pomeroy and Hamilton Oil Company completed its G. W. Besser No. 1 near the Appalachia townsite in the SW ¼, SE ¼, SE ¼ of Section 24-T20N-R9E on March 19, 1919. Flowing 50 barrels daily, the well opened the Keystone Pool to development. Eventually the field was expanded to cover the eastern one third of T20N-R9E and the western quarter of T20N-R10E, and southward into Creek County. However, most of Keystone's production was found in Payne County, where 104 of the field's 178 successful wells were completed.[15]

Southwest of Cleveland, the Maramec Pool was located in 1920. Eventually its productive acreage spread over sections 4, 5, 8, 9, 16, 17, 21, 22, 27, and 28 T20N-R6E. Although some of the early wells produced only natural gas, the Republic Oil and Pipe Line Company brought in a 200 barrel-per-day oil well in Section 21.[16]

There were three offshoots of the original Maramec discovery. Maramec North was located in 1945 in T21N-R6E. However, it proved to be a small find, and by 1956 only one well was producing in the pool. The two other satellite fields were uncovered in 1948: Maramec East in Section 1-T20N-R6E and Maramec Southeast in Section 31-T20-R7E.[17]

Also in 1920, the Blackburn East Pool was located northwest of the original Cleveland discovery in a bend in the Arkansas River in Section 21-T22N-R7E. Two auxiliary fields, Blackburn Southeast and Blackburn South, were found in 1954 and 1955, respectively, southwest of the original discovery. However, only one producing well was operating in the Blackburn Southeast Field two years after the pool's discovery.[18]

Another find, the Garr Field, was made in 1920 in southwestern Pawnee County along its border with Payne County. Of the pool's 73 producing wells only 30 were located in sections 32 and 33 T20N-R5E of Pawnee County. The remainder were in Payne County.[19]

In December of 1925, J. L. Donahue and others began development of the Masham or Donahue Field when the No. 1 Zoldoski was completed in the NW ¼, SE ¼ of Section 22-T23N-R4E. Previously J. Alex Dingwall had completed a 5,500,000 cubic-feet-per-day gasser a mile south of Donahue's well, but found no oil. The No. 1 Zoldoski, on the other hand, flowed 730 barrels daily. Even so, it was not until the 1950s, with the discovery of the Masham East Pool in Section 20-T23N-R5E, that the region was extensively developed.[20]

Several other minor strikes occurred in Pawnee County after the discovery at Cleveland. A show of oil was found in sections 31 and 32 of T21N-R8E and Section 5-T20N-R8E in what was called the Adamson Pool. The Southwestern Field in sections 13, 23, and 24 of T20N-R8E found both oil and gas producing formations. The Kelly-Chandler No. 1, located in the NE ¼, SW ¼ of Section 1-T21N-R5E, showed a sign of oil when it was swabbed on August 27, 1924. Finally the Devonian Pool was uncovered by the Devonian Oil Co. on the W. M. Longmire farm in the SE ¼, NW ¼ of Section 35-T22N-R6E on May 6, 1926; however, its initial production was reported as only 40 barrels daily.[21]

Of more significance was the activity of the Minnehoma Oil Company in the SW ¼, SE ¼, SW ¼ of Section 13-T21n-R7E. The discovery well, the No. 3-W Merrall, was completed on June 30, 1926, and flowed 75 barrels of oil daily from the Hominy Sand between 2,630 and 2,725 feet. Named the Minnehoma Field, production quickly was expanded south into Section 24.[22]

The Minnehoma discovery was but a part of the firm's early involvement in the greater Cleveland area. Formed in 1903, a year before the strike at Cleveland, by George F. Getty, the father of J. Paul Getty, the Minnehoma was one of the major developers of the region and, as a result, eventually extended its influence far beyond the boundaries of Oklahoma. The elder Getty, a successful Minneapolis, Minnesota, attorney, had traveled to Bartlesville in 1903 on business. While there he became intrigued by the nearby oil play, and purchased Lot 50, a 1,100 acre lease in the neighboring Osage Nation, in what became Osage County, Oklahoma. Because it was two miles from the nearest proved production, Lot 50 cost the elder Getty a mere $500.[23]

To handle his investment, Getty formed the Minnehoma Oil Company on December 7, 1903. Although it was organized in Phoenix, Arizona, its name was derived from a combination of the first five letters of Minnesota and the last four letters of Oklahoma. Getty, the principal stockholder and president of the company, was joined by W. F. Bechtel, W. A. Kerr, Frederick V. Brown,

Two different sizes of wooden storage tanks surround this group of wells near Cleveland. Because it was difficult to keep them from leaking, a single large wooden tank usually could not handle a prolific well's production. Note the rig builder working on the rig in the right foreground. (Robert Jordan).

Bill and Sherman Copeland standing beside a wagon designed to haul oil-field pipe in the Cleveland area, ca. 1904. Instead of having a wagon box, the load of pipe was supported by the frame between the wheels. The reins to the team were supported by the rings over the front wheels, thus enabling the teamster to walk beside the wagon. (Triangle Heritage Association of the Cherokee Strip).

Workers building a battery of wooden tanks near Cleveland. Note the stairway leading to the top of the tanks and the string of lights above. This allowed the tanks to be checked 24 hours a day. (Triangle Heritage Association of the Cherokee Strip).

A wooden derrick toppled by high winds near Cleveland. Fortunately for the driller, it appears, from the stack of casing to the right of the derrick, that the well had been completed before the windstorm. (Triangle Heritage Association of the Cherokee Strip).

A sawmill cutting rig timbers near Cleveland during the oil boom. Because the area contained numerous trees, most of the beams needed for rig construction were sawed locally. The large log being trimmed likely would have been squared and used to support the derrick floor. (Robert Jordan).

A pipeline crew laying pipe in the Cleveland area. Because of its centralized location, Cleveland became a major transfer and shipping point for most of the crude produced nearby. Four of the handles of the large tongs used to screw sections of the pipe together can be seen. (Robert Jordan).

John W. Bell, and P. D. McConnel. J. Paul Getty entered the company the following year, 1904, when he purchased 100 shares of stock at a cost of $5.00. This was the beginning of what became the Getty Oil Company.[24]

Eventually 43 wells were drilled on Lot 50. Of the total, 1 was a duster and the other 42 were producers. As a result, almost overnight Minnehoma became one of the largest producers in the region.[25]

On January 19, 1904, shortly after the completion of the first well, George F. Getty, his wife, Sarah MacPherson Risher Getty, and J. Paul, journeyed from Minnesota, where they continued to maintain their home, to Indian Territory to inspect Lot 50. After a three-day train trip, they arrived on January 21, with the younger Getty expecting to find "war-painted Indians, gunslinging cowboys and galloping cavalrymen." He was disappointed when Bartlesville proved to be nothing but a "primitive boomtown, with mainly clapboard buildings—a few false-fronted to give the impression of being two-storeyed—and sagging wooden sidewalks," filled with "oilfield workers and not cowboys."[26]

Although disappointed at the "Wild West," J. Paul Getty was fascinated by the oil business. On January 30, 1904, the family rented a horse and buggy and drove the nine miles from town to lot 50. The younger Getty later recalled that it was a rough ride over unimproved land "down a track through the woods [and] across Sand Creek." Facinated by what he saw, J. Paul spent as much of his time as possible on Lot 50, and thus was present on March 2, 1904, when the Minnehoma No. 2 was completed at 1,426 feet.[27]

Two days later he left with his parents to return to Minnesota. That fall, in October of 1904, the Gettys returned to Bartlesville for another 10-day visit. Again spending much of the time on the Osage lease, on October 17, the younger Getty saw the Minnehoma No. 3 shot with "100 quarts of nitroglycerine." Watching crude gushing over the derrick after the shot, the younger Getty was, as he said, captivated by the "lure of oil."[28]

Afterward the Getty family returned to Minnesota, and then moved to Los Angeles, California. Even so, the younger Getty returned to Oklahoma as often as possible. In the intervening years, Minnehoma Oil Company continued to grow, as did J. Paul's interest in the firm. In 1909, when he was 16 years old, he asked his father if he could spend his summer vacation working on Minnehoma's Osage lease. Agreeing, his father stipulated that he would have to be "willing to start at the bottom" and would receive "no preferential treatment because...[he] was the Boss's son." Instead he would "have to hold...[his] own with the other men, take...[his] share of the orders and do...[his] share of the work."[29]

Big for his age, at 16 the younger Getty stood almost six feet tall, weighed "a fat-free eleven stone five [159 pounds] and [was] in excellent physical condition." Because of his size, he was made a roustabout and given the "heaviest (and usually the dirtiest) work on a drilling site." Most of his daily 12-hour tour was spent unloading, assembling, shifting, and operating machinery; stacking lengths of pipe; and laying pipelines from well heads to storage tanks. "All in all," he continued, "there were a thousand and one chores, not a few of which called for rapid, monkey-like scrambles up and down the seventy-four-foot derrick or other forms of extreme physical exertion." His pay, like that of the other roustabouts, was $3.00 a day.[30]

Working "without a complaint," at least "after the first week or so, when the blisters on my palms had begun to toughen into protective callouses," the younger Getty, because of his hair, was quickly nicknamed "Red" by his cohorts. Although only 16, he joined his co-workers on their "payday-night donnybrooks" in the nearby oil boom towns, where, as he recalled, bootleggers "palmed off" muriatic acid, "on the thirst-parched as whiskey." It did not take Getty long to learn the roughneck's trade, and afterward, because he was there to learn all he could about drilling a well, he was assigned to a longtime oilman, who had the nickname of "Grizzle," to learn to be a tool dresser or "toolie" as they were commonly called.[31]

It was one of the most important jobs on a cable tool rig. Assigned the responsibility of assisting the driller and insuring that the drill bits always were sharp and in working order, "the toolie had to be a crack technician and an expert, highly specialized blacksmith." As Getty explained, the success of the well often depended on the toolie, and "the quality of his work at his rigside forge and anvil could make or break vital drilling bits—and in time, even the entire drilling operation." By the summer of 1911, Getty had acquired enough experience to be assigned as a regular tool

Looking across the Arkansas River toward Cleveland. The smoke in the far background is coming from a burning well in the Boston Pool across the river in the Osage Reservation. (Triangle Heritage Association of the Cherokee Outlet).

Stout's Ferry, across the Arkansas River northeast of Cleveland, connected the community to the Boston Pool and Osage City. (Robert Jordan).

A wooden foot bridge built over the Arkansas River so that Cleveland oilmen could more easily reach the Boston Field. (Triangle Heritage Association of the Cherokee Outlet).

dresser on one of the Minnehoma Oil Company's drilling crews.[32]

Because Minnehoma's leases in Osage County were due to expire in March of 1916, as early as 1911 George F. Getty began looking at the Cleveland area for replacement sites. Impressed with the activity to the south of the Osage, by 1913 he had purchased 7 leases around the town of Cleveland. This action, according to J. Paul Getty, "assured the future of Minnehoma."[33]

Among Minnehoma's Cleveland properties were three 80-acre leases; the Baker farm, acquired on February 27, 1911; the Michaels farm, acquired on March 1, 1911; and the Russell farm, of which no acquisition date was recorded. Because they were two and three miles, respectively, from the nearest production, the Baker and Russell leases were relatively cheap. However, the Michaels lease was only 3/8 of a mile from a proven well, and cost the company $11,050.[34]

The two extremes were developed first, the Russell farm, which was the most distant from proven production, and the Michaels lease, which was closest to an already producing well. Drilled in 1911, both proved to be extremely prolific, and during the next few years produced in excess of 1 million barrels of crude, which netted an $850,000 profit to Minnehoma. The Baker lease was drilled in 1912, and although it did not prove to be as prolific as the Russell and Michaels property, it produced a total of 90,170.22 barrels of crude before it was abandoned in 1937.[35]

Encouraged by the initial success at Cleveland, Minnehoma purchased two other acreages in the pool in 1913: the McFall lease, which contained 40 acres, and the 640-acre Section 31 Carson lease. The McFall property was drilled almost immediately and by 1937 had produced a total of 6,950,781 barrels of oil. The Section 31 Carson, located to the south of Cleveland, proved to be the least productive of the seven leases, and between 1913 and 1922 produced only 172,355 barrels of crude. It was abandoned in 1922.[36]

When the huge Cushing-Drumright Field was opened in northwest Creek County, just to the south of Pawnee County, George F. Getty wasted little time in becoming involved. In May, 1914, Minnehoma purchased 5/8 interest of the Lete Kolvin allotment, just north of the strike in Section 16-T18N-R7E, from L. L. Hutchinson, P. G. Walker, Walter E. Templeman, and F. L. Warren for $21,875. The value of the property skyrocketed, and in a little more than two years Minnehoma sold a 1/8th interest in the lease to John T. Milliken, President of Milliken Oil Company, for $17,500.[37]

Eventually an iron bridge was built across the Arkansas River to connect Cleveland with the wells in what became Osage County. However, in 1924 the bridge, which can be seen in the left background of this photograph, was destroyed by a flood and a new ferry was opened to handle the traffic. Clayton Lucas is the man on the right pulling on the ropes holding the ferry to the shore. (Triangle Heritage Association of the Cherokee Outlet).

The Osage Inn in Osage City, located to the east of Cleveland across the Arkansas River. It became one of the most famous restaurants in the area. In this snow covered scene uniformed waitresses and cooks are posing on the porch of the Inn, along with two horses. (Robert Jordan).

George F. Getty, who, along with his son J. Paul Getty, laid the foundation of what became Getty Oil by establishing the Minnehoma Oil Company to develop several leases in the Cleveland area. (PennWell Publishing Company).

During this period of Minnehoma's expansion in Pawnee and Creek counties, J. Paul Getty continued to work in the Oklahoma oil fields during his summers until 1914 when he travelled to Europe to complete his education. However, shortly thereafter, in September, 1914, he was forced to return to the United States because of the outbreak of World War I. At that time, Getty accepted an offer by his father to become an independent operator in the Oklahoma oil fields.[38]

In exchange for a salary of $100 a month the younger Getty "scouted for low-cost leases on properties that either instinct or my knowledge of surface geology told me might bear oil." Once he found a lease, his father, for a 70 percent share, furnished the money to develop the property. J. Paul retained ownership of 30 percent of any leases he uncovered.[39]

Working out of a $6.00-a-week room in Tulsa's Cordova Hotel, with another $6.00 weekly budgeted for meals at a nearby boarding house, J. Paul Getty spent most of his time "scouring the countryside in a third-hand Model T. Ford." Although the oil boom had "insanely inflated" the price of most leases, he managed to acquire the Nancy Taylor allotment in Muskogee County for "the astoundingly low price of $500." The first well on the lease produced 700 barrels daily, and the lease proved to be so profitable to the Minnehoma Oil Company that J. Paul Getty was named one of the firm's directors.[40]

After the Taylor success, the younger Getty obtained additional properties and by mid-1916, "was a millionaire." In May of that same year, George F. Getty and J. Paul Getty incorporated the Getty Oil Company, with the elder Getty serving as president and treasurer of the firm, and his son the secretary. As in their other business arrangements, George Getty received 70 percent of the stock and J. Paul Getty 30 percent.[41]

Also in 1916, J. Paul Getty ended his joint ventures with his father and the Minnehoma Oil Company and "undertook independent enterprises," which, as he later explained, "I financed entirely with my own private capital." Seven years later, in 1923, he inherited the leadership of the Minnehoma Oil and Gas Company and George F. Getty, Inc. Because his father's and his own interest were "primarily engaged in oil exploration and production," he combined these firms with Getty Oil Company and laid the ground work for a multinational corporation that by 1974 included a multitude of corporations spread over four continents.[42]

After the activity of the mid-1920s no new fields were uncovered in Pawnee County for a number of years. The huge strikes at Seminole and Oklahoma City, together with the tremendous production of East Texas, flooded the market and so depressed the price of crude in the Mid-Continent Region that additional wildcat activity was greatly curtailed. It was not until the outbreak of fighting in Europe in 1939 and the resulting rearmament program in America that the nation's economy supported a revival of the oil business in the area.[43]

George Getty's leases at Cleveland "assured the future" of what eventually became Getty Oil and during the following decades the corporation became one of the largest energy concerns in the world. By the early-1980s, numerous Getty service stations offered their products not only across the nation, but throughout Oklahoma. (Getty Refining and Marketing Company).

The first strike of Pawnee County's second oil boom was made in 1940, just to the west of the Keystone Field in Section 28-T20R-R9E. Named the Mannford North Field, it proved to be a short-lived, small producer. The next year, 1941, another small field, the Pawnee East Pool was located in the northwestern corner of T21N-R5E. However, it was abandoned in 1944 after producing only 8,731 barrels of oil.[44]

It was not until 1942 that another major find was made in Pawnee County. In that year, just to the southwest of the original Uncle Bill No. 1, in the southeastern part of T22N-R6E, the Skedee Field was uncovered. Two other nearby discoveries were made because of the increased drilling activity resulting from the Skedee strike: the Skedee West, in Section 15-T22N-R5E, and the Skedee North in sections 22 and 23 of T22N-R6E.[45]

In the following decade, the 1950s, a large conglomeration of pools was found in southwestern Pawnee County. The Bryan North was located in 1951 in sections 26 and 35 T21N-R5E. Two years later, in 1953, the Bryan Northwest and Bryan Southeast were found in Section 33-T21N-R5E and Section 12-T20N-R5E respectively. The following year the Bryan Pool was found in sections 1 and 2 T20N-R5E and the Bryan Northeast was located in Section 1-T20N-R5E.[46]

Several additional strikes were made in 1954. The Cody Pool was found in the extreme northwest corner of T21N-R6E. Eventually nine wells were drilled within the area. A few miles west of the Arkansas River, in sections 27 and 34 of T23N-R5E, the Crescent Star South was discovered. In 1955, the Cody Southwest was opened in Section 11-T21N-R5E, and within a year the discovery well had produced 1,896 barrels of oil. That same year the Sunny Slope Southwest Gas Field was located just to the northeast of the original Cody strike.[47]

Although the discoveries in Pawnee County following the completion of the Uncle Bill Lowery No. 1 did not reach the magnitude of those made in Payne, Carter, Seminole, or Oklahoma counties, Pawnee County did maintain a significant production of oil and natural gas through the years. At the end of 1980, Pawnee County's annual output of crude stood at 1,064,454 barrels. This was enough to rank the county 30th

Frank Buttram, a pioneer Oklahoma oilman who opened the Watchorn or Morrison Field in northwestern Pawnee County. (Oklahoma Heritage Association).

A bridge crew operating a pile driver while constructing a bridge over the Arkansas River near the Blackburn pools northwest of Cleveland. (Triangle Heritage Association of the Cherokee Outlet).

among the state's 77 counties in oil production. That same year, Pawnee County's production of natural gas reached 620,029,000 cubic feet. By January of 1983, Pawnee County's monthly output reached 122,558 barrels of oil per month, 28th in rank among Oklahoma's oil-producing counties.[48]

Although the oil production of Pawnee County was relatively small, it served to maintain the interest of oilmen in the region at a time when the federal government's policy of leasing Indian land at Red Fork had tempted many pioneer oilmen to abandon the area. In addition, it served as the harbinger of things to come. For within a year of the completion of the Uncle Bill No. 1, the discovery of crude at Glenn Pool ushered in the heyday of Oklahoma's petroleum industry.[49]

A 55,000 barrel oil storage tank burning on the A. H. Barnes farm outside of Cleveland. The men posing with shovels have just completed repairing the earthen dike surrounding the tank to prevent the burning oil from spreading. (Triangle Heritage Association of the Cherokee Outlet).

The Toxway Oil Company's earthen storage pit near Cleveland. Although such pits were a cheap and quick way to store excess crude, much of the oil was lost through seepage or evaporation. (Triangle Heritage Association of the Cherokee Outlet).

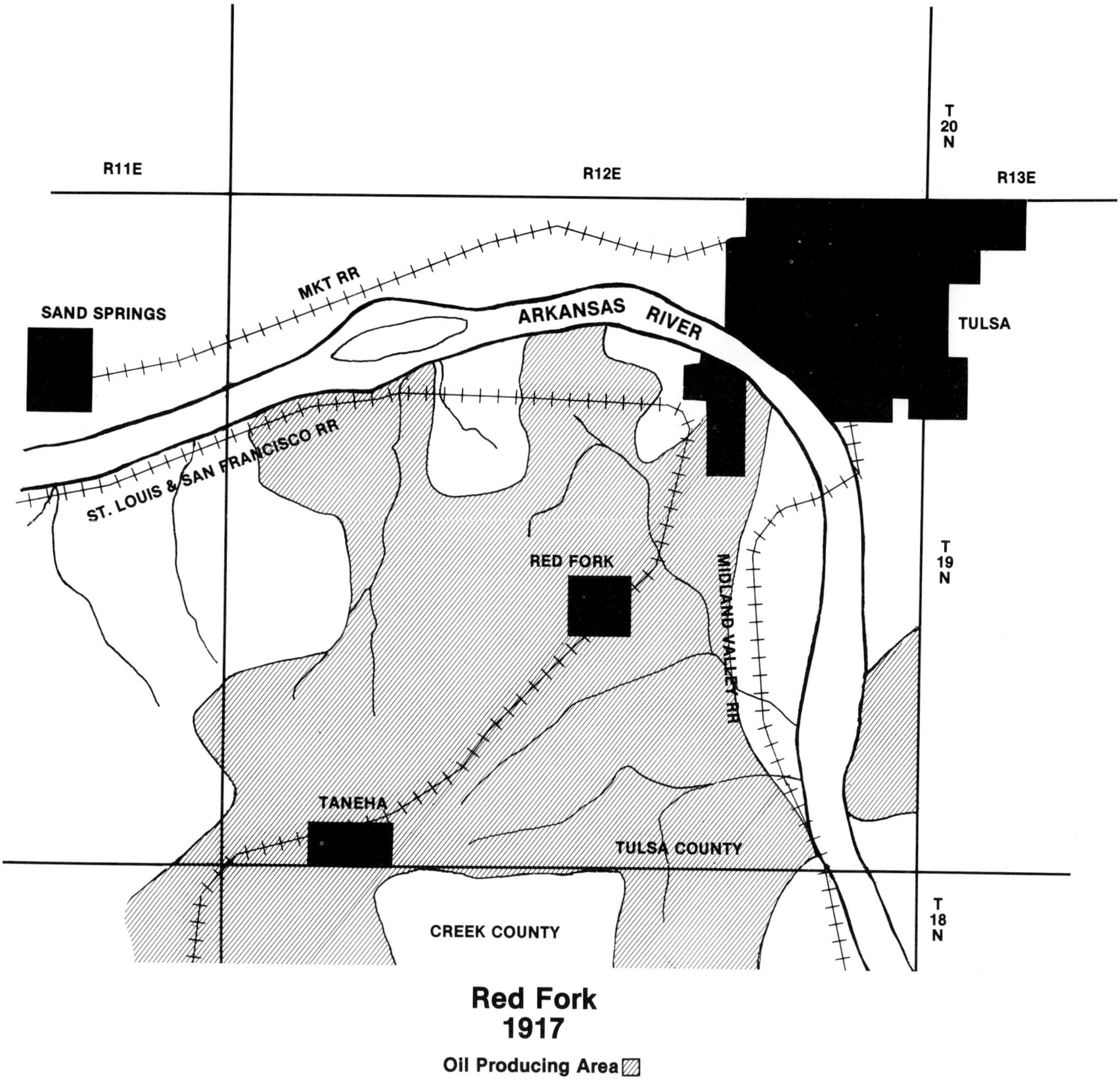

Red Fork, 1917. (Oklahoma Heritage Association)

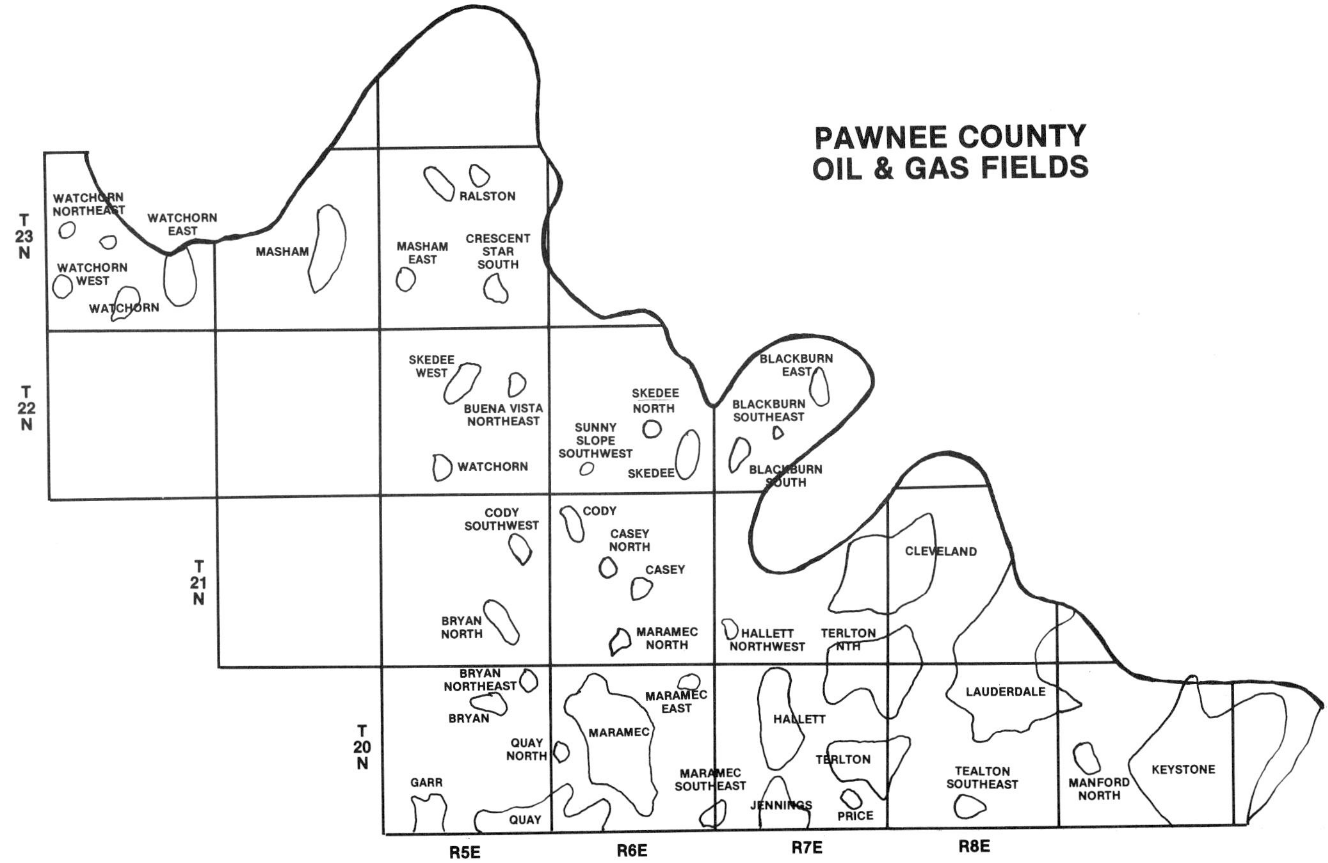

Pawnee County Oil and Gas Fields. (Oklahoma Heritage Association).

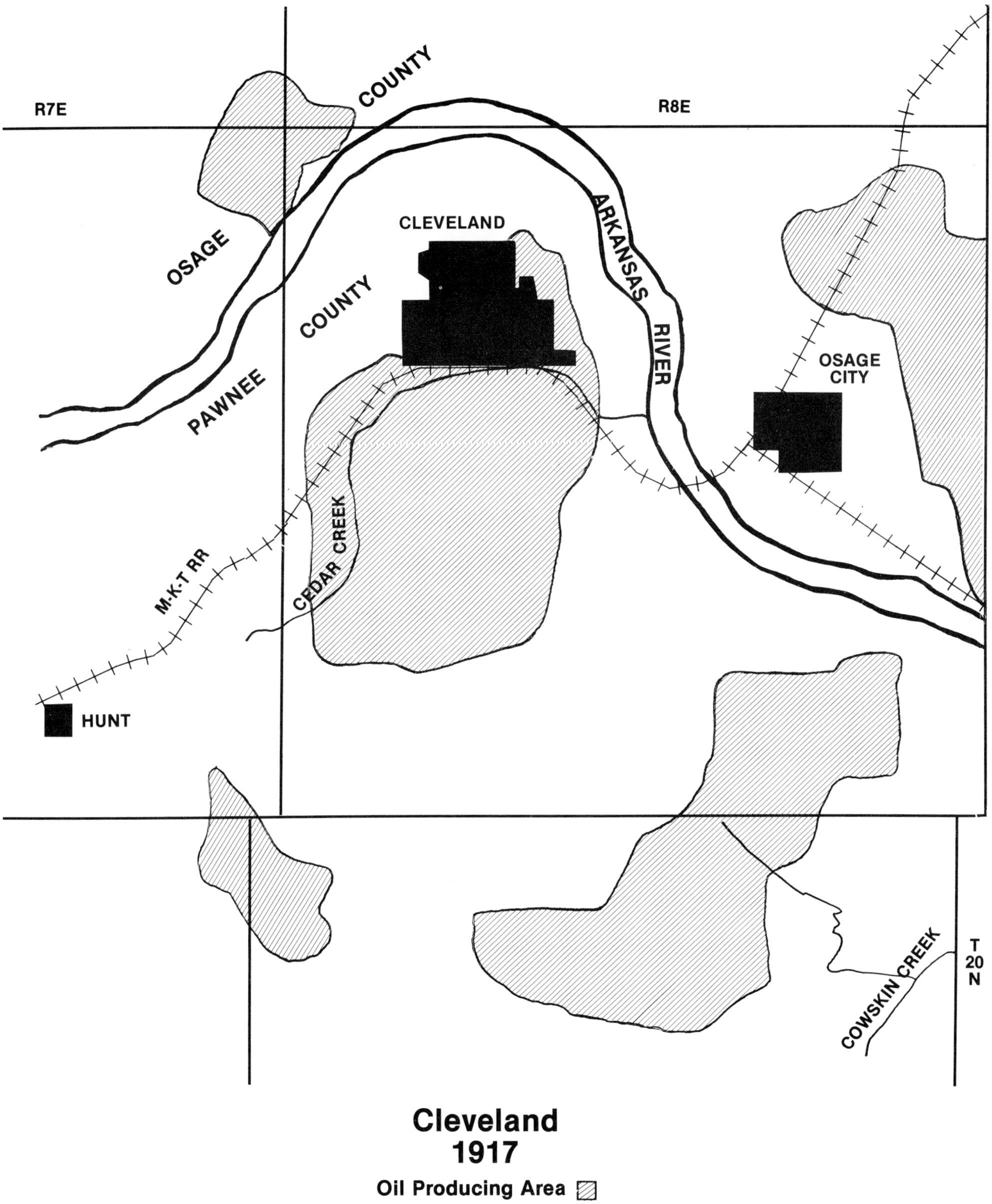

Cleveland, 1917. (Oklahoma Heritage Association).

CHAPTER SIX

THE IDA GLENN NO. 1

One pioneering oilman involved in the strikes at Red Fork and Cleveland was Robert Galbreath, who later launched the great Oklahoma oil boom with his own discovery at Glenn Pool. Born in Packaway, Ohio, Galbreath had moved to Indian Territory in 1889 when the Unnassigned Lands were opened to homesteaders. Although he succeeded in claiming 160 acres in Turkey Creek bottom near Hennessey, he soon sold the property and moved to Edmond where he became the community's first marshal. Later Galbreath participated in several other Oklahoma land openings and eventually was appointed a United States Commissioner, first at Perry and then Shawnee. By 1899 Galbreath had moved to Oklahoma City and opened a real estate business. As soon as he learned of the Red Fork strike, however, he hurried to the Creek Nation community.[1]

"I'd never seen an oil well before, but the novelty of the thing attracted me," Galbreath admitted. "It seemed to give me a wonderful enthusiasm to tackle that business...and I had visions of making a lot of money," he continued. By investing the money he had received from selling his previous land claims, Galbreath planned "to take a 'shot' at the oil business." When he confided his plans to his friends, many were convinced it was a "foolish" decision."[2]

However, he did persuade Charles Colcord and C. G. Jones to support him financially, and with their backing Galbreath acquired several Red Fork leases. Ignoring the regulations requiring federal approval of his leases, Galbreath wasted little time in starting his wells. His investment proved to be profitable. One Galbreath well, the Nathaniel Yargee No. 3, located about a mile northeast of the Sue A. Bland No. 1, was completed as a 125-barrels-per-day-producer, and another on the Van Yargee lease, which later became a part of Howard Park in Tulsa, produced 150 barrels daily. Eventually, Galbreath's Red Fork wells prompted him to move his family to the community, where he rented a house from Louise Clinton.[3]

Like many other oilmen, Galbreath began to expand outward from the original strike in search of other nearby oil producing formations. To the southeast of Red Fork was a series of broken surface formations that attracted his attention, and in early July, 1901, he began examining the area along with another local resident. On the night of July 3-4, Galbreath and his friend camped for the night near a horse lot on Eck E. Brock's 320-acre farm southeast of Red Fork. After discussing their findings, Galbreath decided to visit the the nearby Ida E. Glenn allotment, a mile north of the Brock farm in the center of Section 10-T17N-R12E, the following morning. A one-eighth Creek, Ida Glenn farmed her allotment with her husband Robert.[4]

When he arrived at the Glenn farm and explained his reason for examining the region, Galbreath was astounded to learn that the Glenns knew of a nearby limestone ledge that was "heavily stained" with crude. Persuading the Glenns to show him the site, Galbreath was more excited when, after breaking off a piece of the ledge, it yielded a "module of thick green oil." Although the Glenns explained that "everyone for miles around" knew about the ledge and nearby oil seeps, Galbreath made them promise to keep his visit a secret. In exchange, he pledged that as soon as the Department of the Interior eased the regulations governing the drilling of wells on Indian land, he would sink a wildcat well on their farm.[5]

Preoccupied with his Red Fork leases, it was not until early in 1904 that Galbreath returned to the Glenn farm. True to their promise, the Glenns had not leased the property to anyone else. Because he was short of money, Galbreath invited William S. Guthrie and Weston Atwood to visit the oil-soaked ledge in hopes that they would invest in the project. Satisfied that the area was a potential petroleum producer, Guthrie and Atwood agreed to join the venture. With adequate backing, Galbreath wasted little time in signing a lease for the Glenn property. In exchange for $45, approximately 3¢ per acre, and a ⅛th royalty on any oil or natural gas found, the Glenns gave Galbreath the exclusive right to drill.[6]

In addition to the Glenn lease, Galbreath also signed agreements with Elizabeth Crosby, on April

The Ida Glenn No. 1, discovery well of Glenn Pool. (Oklahoma Historical Society).

Nitroglycerin was widely used at Glenn Pool to stimulate production. The photographer caught the plume of oil being thrown into the air just as this well was shot with 160 gallons of nitroglycerin. Note the steel derrick, indicating that this well was drilled during the 1920s or later. (Sapulpa Historical Museum).

19, 1904, and Robert Barver, on April 22, 1904. Crosby owned 160 acres in Section 2-T17N-R12E, about a mile north of the Glenn farm, and Barver held a quarter-section in Section 22-T17N-R12E, two miles to the south of the Glenn allotment. Although Galbreath's Glenn lease was approved by the Department of the Interior on September 22, 1904, the other two leases were not approved by federal officials until May 9, 1905 and April 4, 1906 respectively.[7]

Ready to began operations, Galbreath received quite a setback when Guthrie and Atwood suddenly withdrew their financial support. "My various partners ran hot, then cold, on the idea of the well down in the Mounds country," Galbreath explained. Unperturbed, Galbreath decided to go it alone and "bought them out."[8]

Not only did Guthrie and Atwood have second thoughts about sinking a wildcat well on the Glenn farm, Galbreath continued, "even Roy Dodd, my partner in the drilling contracting business at Red Fork, took a dim view of the deal." When Galbreath offered Dodd "a share of the lease for the use of one half of the drilling tools," Dodd replied " 'Bob [Galbreath] I'm not too interested in that lease, but I'll sell you my half share of the tools for $1500." Protesting, Galbreath pointed out "you know damn well I haven't got $1500 to spare or I'd be drilling that well by now by myself." Dodd then offered to let Galbreath "take the drilling tools off my hands and pay me when you can." He also agreed to "work...on this rig as a driller for $5.00 a day." Galbreath quickly accepted.[9]

With Guthrie and Atwood out of the picture, Galbreath attempted to lure other local investors into the project. Although Bates Burnett, a Sapulpa banker, apparently agreed to advance some

Shooting a well was a dangerous business. This photograph shows a hole made when a nitroglycerin magazine, belonging to the Central Torpedo Company, exploded unexpectedly on March 30, 1912 near Glenn Pool. The depression is 25 feet across and 15 feet deep. A witness to the explosion reportedly heard hammering just prior to the blast. Although the exact cause could not be determined, the blast probably was touched off by the hammering. (PennWell Publishing Company).

money, Galbreath found few other investors who were interested in backing such a "risky" proposal. Galbreath finally turned to fellow Tulsan, Frank Chesley.[10]

Chesley, one of the founders of the town of Keystone just to the west of Tulsa at the junction of the Arkansas and Cimarron rivers, was a well-known area oilman and according to Galbreath "the best field man I ever knew." However, Chesley remained unconvinced of the possibilities of the Glenn property until Galbreath showed him the limestone outcropping. After breaking off a piece of the ledge to see the crude for himself, Chesley readily agreed to underwrite the well.[11]

With Dodd's and Chelsey's help assured, Galbreath wasted little time in securing a drilling crew. "Shorty" Miller, one of the region's best known drillers, was hired to oversee the operation. Located near the center of Section 10-T17N-R12E, about four miles southeast of Sapulpa, the Ida Glenn No. 1 was spudded in during the fall of 1905. Although several nearby strikes had been made, most local residents had never seen a well drilled and the activity attracted a great number of sightseers. At first, Galbreath made no attempt to keep the operation a secret, and "inhabitants for miles around, including the cowboys from the ranches, would visit the operations daily and watch with wonder the drillers work."[12]

Because of the shortage of cash, which forced the well to be drilled without casing, Galbreath and Chelsey lived on the drilling rig along with

The derrick for this Glenn Pool well was removed prior to the well being shot with nitroglycerin, however, for some reason the walking beam and other equipment was left intact. (Oklahoma Historical Society).

A typical cable tool rig in operation in Glenn Pool. This view clearly shows the the iron stirrup which connects the walking beam to the band wheel. As the band wheel is turned by the boiler, the stirrup transfers the motion to the walking beam, which raises and lowers the cable supporting the bit in the hole. (Sapulpa Historical Museum).

Dodd and Miller. Not merely observers, Galbreath and Chelsey worked with the drilling crew to keep the boilers going. "We used coal as fuel for the boilers," Galbreath explained, and we "took our turns in firing to keep a good head of steam." Short of funds, the oilmen often were forced to shut down the drilling rig so they could mine enough coal from a "little hillside mine on an adjoining tract to the south" to keep the fires burning. Nonetheless, by working around-the-clock "we were able to make good time with nothing out of the ordinary interfering with our plan."[13]

Throughout October and November of 1905, they drilled without a trace of petroleum. Suddenly, at the 1,300 foot level, Galbreath, in a ploy to discourage sightseers from keeping their vigil on the operation, announced that the project was being abandoned. Galbreath was afraid that in the event of a gusher, too many unwanted witnesses would set off a wild rush to the area before he had the opportunity to secure nearby leases.[14]

However, when the drill bit passed 1,400 feet without encountering any sign of oil even Galbreath became discouraged. The gamble appeared lost, and on the evening of November 21, 1905, Galbreath, Chesley, and the others discussed the possibility of actually abandoning the hole. Reluctant to give up, they decided to continue to 1,500 feet before making a decision.[15]

Their persistence was rewarded, for near dawn the next day, October 22, 1905, just after Chesley had replaced Galbreath on the derrick floor, Chesley noticed a stain on the bit. He immediately called to Dodd to haul the tools to the surface and lower the bailer to take a sample. When the operation was completed and the bailer pulled back to the surface, Chesley took one look at its contents and rushed to awaken Galbreath. Yelling "Oil!

A side view of the stirrup and walking beam on a cable tool rig at Glenn Pool. The walking beam rests on top of the sampson post, against which a crude ladder is leaning. The large wheel to the right of the derrick is the bull wheel, around which is wound the cable that supports the bit in the hole. The wheels at the top of the derrick, called the crown block, are the sand pump pulley, on the left; the casing pulley, in the center; and the crown pulley, on the right. (Continential Oil Company)

Oil! My God Bob...We got an oil well," he shook Galbreath from bed and shoved some oil sand at him.[16]

Galbreath looked at the sample Chelsey was holding and then rushed to the well head. By the time he reached the hole, it was approximately 5:00 a.m., and the crew could hear a noise coming from below. Galbreath described it as "quietly chuckling and surging and slooping." Just as he leaned over to look down the hole, the Ida Glenn No. 1 blew in. "We watched in wonderment," Galbreath explained, as the crude "mounted from a quiet gaseous stream into an increasingly forceful column of golden riches that rapidly mounted higher and higher and higher." "Then," according to Galbreath, "all hell broke loose, and the Ida E. Glenn flowed mightily over the derrick top, over the crown block, high into the air above." "We had gambled and won," he declared, as the well was completed as a 75-barrels-per-day producer.[17]

Despite Galbreath's efforts to minimize the significance of the new strike, on November 23, 1905, the day after the well blew in, the *Tulsa Democrat* published a rumor that Galbreath and Chesley had made a new discovery. The story was expanded the following day when the paper reported that if Galbreath and Chelsey had indeed made a strike it would surely "open up a big stretch of territory" to oil development. The actual completion of the well was reported by the *Mounds Enterprise*, on November 24. Its story reported "a strong flow [of oil] is struck, eight miles from town...on Bob Glenn's place."[18]

As early as November 26, several Tulsa oilmen attempted to visit the Ida Glenn No. 1 and see for themselves if a new field had been found. By this time, however, Galbreath was taking no chances in maintaining absolute secrecy, and as they "reached a point within a few hundred feet of the well" on a road that wound through a corn field, "a guard appeared, gun in hand" and declared that the road ended there. Although the oilmen did not see the well, the armed guards seemed to verify that a major find had been made. With this as confirmation, operators from throughout the territory and surrounding states rushed to the area.[19]

Still short of money, once the Ida Glenn No. 1 was completed Galbreath again approached Colcord for financial backing. This time Colcord readily agreed and gave Galbreath $2,500. He eventually received a return of nearly $400,000 for his investment. Likewise, Galbreath persuaded John O. Mitchell, a Tulsa attorney, to advance him $1,000. The four men—Galbreath, Chesley, Colcord, and Mitchell—quickly formed the Creek Oil Company.[20]

Ironically, Galbreath was unable to cash Mitchell's note in any Tulsa bank. When he retuned to Red Fork, however, he found a banker who cashed the check and asked him "why he had not come to him in the first place." All Galbreath could answer was that "he thought all the money was in Tulsa."[21]

Although he now had adequate financial backing, it was four months before casing was set in the Ida Glenn No. 1. In the meantime, Galbreath and his partners embarked on a campaign to acquire additional property in the area. Within

A horse-powered pipe-pulling rig operating in Glenn Pool. Mounted on wheels, the contraption was pulled to the well site by a team of horses, set up, and then the same team of horses were used to extract the pipe from the hole. (Sapulpa Historical Museum).

a short time the Creek Oil Company held leases on the Eck Brook farm, a mile south of the discovery well, as well as the nearby J. E. Morrison, Mable Glenn, and Gracie Glenn allotments.[22]

Leases in the area previously had been obtained for as little as $2.00 per acre, but with the discovery of oil the price shot up. One lease broker reported that $2,500,000 had been offered and rejected for a promising location. Another oilman, David F. Connelly of the Oklahoma Associated Producers, a subsidiary of Tide Water Oil Company, paid a $12,000 bonus for the Elijah Corbray farm, one mile south of the Ida Glenn No. 1. Likewise, W. H. Milliken paid $17,000 as a bonus for another nearby lease, owned by Thomas Gilcrease. In one instance Galbreath delivered $30,000 in gold coin in "an old 'gunny' sack" to purchase a promising piece of property. In addition to lease money, fortunate landowners were paid massive sums in royalties—almost $700,000 during the first year of production.[23]

For a time, the Department of the Interior threatened to intervene in the development of Glenn Pool. Galbreath explained that he knew all along that "there was some doubt whether we could get Indian departmental approval on some of the leases." Eventually, however, federal officials gave their approval to all of Galbreath's leases and with that barrier finally removed drilling began in earnest.[24]

Although the Department of the Interior allowed Galbreath to lease Creek land, it did insist on specific conditions. Under departmental regulations, Interior Department officials had to approve all leases on Indian land. In addition, a "straight royalty" of 10 percent was set for all oil sold, with the money paid to the Indian agent at Muskogee, who in turn distributed the funds to

An overview of Glenn Pool. In the center of the photograph, teams and scrapers are preparing a new well site while rig timbers are being unloaded. Note the shotgun house, with the wash hanging on the clothes line, near the well site. In the foreground are covered wagons used to transport perishable supplies. (Oklahoma Historical Society).

The drilling equipment on this cable tool rig at Glenn Pool has been removed and a pumping unit installed. The same walking beam that raised and lowered the cable in the hole has been adapted to raise and lower the plunger in the pump. (Sapulpa Historical Museum).

the proper Indian landowner. While the welfare of the Indians in dealing with oilmen was scrutized by federal officials, agreements between oilmen and freedmen (former slaves held by the Indians) were not. The freedmen held deeds to their property and had the right to dispose of their property however they desired.[25]

Unfortunately, many were taken in by unscrupulous spectulators. For example, in early 1906 a farmer from Mounds, who happened to be in Muskogee on business, paused to visit with a local attorney while he was waiting on the return train. Soon the farmer and the attorney were talking about the discovery at Glenn Pool, and eventually the attorney asked the farmer if he knew of any unallotted Indian land near the strike. After thinking a minute, the farmer recalled that there was still some unclaimed land near the Ida Glenn No. 1. The attorney then confided to the farmer that he was representing a Creek freedman who, although in jail at the time, was still entitled to an allotment. All that was need to free the black from jail was $800.[26]

Within a short time the farmer and the attorney reached an agreement whereby they would put up $400 each to free the black man from jail if he would agree to assign all of his allotment except a 40-acre homestead, which he was required by law to retain, to them. Once the money was

A Gypsy Oil and Gas Company wagon, pulled by 20 horses, hauling a boiler through Kiefer to a lease in Glenn Pool. Such multiple hitch arrangements were necessary to pull the heavy equipment over the field's primitive roads. (Sapulpa Historical Museum).

raised, the two partners approached the jailed man. Would he, they asked, be willing to file for the Glenn Pool allotment, keep 40 acres as his homestead, and sign the remaining 120 acres over to the attorney and the farmer? He quickly agreed and signed away a potentially valuable lease.[27]

After the discovery at Glenn Pool, many Creeks who had been struggling to make a living from their lands found themselves instantly wealthy. Many of the allottees in the region of the Glenn farm were minors, and the leasing of their allotments was conducted by the courts. Any money derived from the sale was held in trust until the minor became of legal age. Some of the youngsters acquired huge sums of money. Three-year-old Earl Berryhill, for example, was offered $25,000 for his quarter section, and another, five year old Robert Pitman, Jr., was paid $49,000 plus 10 percent royalty for his 160 acres.[28]

If the court decided that a minor had not been paid a fair price for their land, the judge could invalidate the lease and sell it to someone else. For example, Elna Glenn, who was two years old when the discovery was made, was offered $16,000 for her 20-acre allotment, which was just to the northwest of the Ida Glenn No. 1; however, the courts ruled that the offer was inadequate, and a year later, George W. Barnes of Muskogee paid $43,000 for the same property.[29]

Judge T. C. Sulzbacher handled many of the hearings dealing with the minors' leases. Under his direction the initial offering of their property was made at Sapulpa on March 26, 1906. However, because of allegations that the sale had not been properly advertised, a fact seemingly substantiated by the sale of most of the property below its actual value, Galbreath filed suit in federal court to overturn the results. Successful in his suit, another sale was scheduled for June 14, 1906.[30]

The prices paid for leases varied greatly depending their proximity to proved production. Two allottees who held property near the discov-

A coal-fired boiler in operation in Glenn Pool. Such equipment required constant monitoring, both to keep an adequate supply of coal in the furnace and to maintain the correct pressure to gain maximum power without causing an explosion. (Sapulpa Historical Museum).

ery well, John P. Rhodes and Levy R. Kiefer, received \$27.00 an acre from the Eastern Oil Company and \$17.25 an acre from the Prairie Oil and Gas Company, respectively, for their holdings. However, others whose land was more distant were paid less. For example, Anna Kiefer's property was leased for \$1.55 an acre by F. M. Selby, and Stella F. Kiefer received \$2.50 an acre for her farm.[31]

Much of the money paid to minor Creeks was in the form of bonuses, which were in addition to regular royalty. Bonus money also varied from lease to lease. In June, 1906, the Galbreath Oil and Gas Company paid bonuses ranging from Joseph F. Berryhill's \$12,600 to Lena Posey's \$475. That same month the company paid Edith Durant \$16,800 for a bonus on an 80-acre lease. Such payments pumped a vast amount of cash into Glenn Pool. When the Galbreath Oil and Gas Company's bonuses for June, 1906, were added to the Barnes Investment Company's bonuses for that same month, the total amount of money paid was \$69,860 for 3,200 acres of land.[32]

While the courts established a system of trust funds to protect the children involved, nothing was done to protect adults. The new-found wealth attracted legions of unscrupulous promoters, and many Indians fell victim to these con men. In addition, Indians who previously had disposed of their property were shocked when the new owners reaped huge profits over the original price. One non-Indian had paid \$200 for a lot that after the strike brought \$12,000.[33]

Not all landowners who benefited from Glenn Pool were Indians. For example, Zeke Moore, a black man serving time for horse stealing, emerged from jail to find that royalty from production on his 125 acres was an astounding \$400,000. Of course, the original investors fared well also.[34]

By April, 1907, the land for which Connelly had paid \$12,000 shortly after the completion of the Ida Glenn No. 1 in the fall of 1905, held 21 producing wells which had accounted for 8 million barrels of crude. Likewise, by February of 1906 additional wells on Galbreath's Glenn lease were producing 2,280 barrels of crude daily. Galbreath and Chesley at one time were offered \$1.5 million each for their leases—and they refused the offer.[35]

One of Glenn Pool's most prolific producers was Milliken's Gilcrease lease. During Glenn Pool's heyday, Milliken received \$900,000 annually from the property. Eventually, 40 wells were drilled on that single quarter section, which had produced so much crude by 1909 that 1 million barrels of it was held in storage because it could not be sold.[36]

Faced with federal regulations requiring wells on Indian land to be drilled within 90 days of the approval of leases by the Department of the Interior, Galbreath and his partners in the Creek Oil Company hurriedly spudded in several other wells on the Glenn farm. The Glenn No. 2 was drilled about 300 feet from the discovery well. Completed on March 16, 1906, it flowed 1,700 barrels during its first 24 hours. The oil was accompanied by such a strong gas flow that the crude was thrown over the top of the derrick for an hour after the well blew in. Later, the Glenn No. 2's production dropped to between 700 and 800 barrels per day. In May, 1906, two additional gushers were completed on the Glenn lease as 1,000-barrel-a-day producers.[37]

On the Graves farm, two allotments to the west of the Glenn farm, J. C. Elred and M. H. Mosier drilled another well just as the Glenn No. 2 was being completed. It flowed 75 barrels daily. Shortly afterward, another offset to the discovery well, the

The results of an unattended boiler could be disastrous. This boiler explosion on a Gypsy Oil and Gas Company lease in Glenn Pool on August 25, 1910, destroyed much of the drilling equipment. (Sapulpa Historical Museum).

Ida Glenn No. 3, was brought in as a 1,600 barrel-per-day-producer.[38]

In January, 1906, the Creek Oil Company began development of its Eck Brock lease. Although, the Brock No. 1 was abandoned after the tools were caught in the hole, almost immediately Galbreath spudded in the Brock No. 2. It was completed in April, 1906, and along with the crude was a tremendous flow of natural gas. The escaping gas was ignited by one of the lamps lighting the derrick when the well blew in. The flames quickly spread to the oil soaked rig, which burned to the ground, seriously injuring Frank McGuinn, of the drilling crew.[39]

In spite of the two setbacks, Galbreath quickly started another hole on the Brock lease. On June 21, 1906, the Brock No. 3 was completed as a "natural" with a flow of between 1,000 and 1,500 barrels daily. Plans eventually were made to drill a total of 16 wells on he Brock lease, which proved to be very prolific.[40]

As was the case with many early Oklahoma finds, there was no readily available means to market Glenn Pool's crude. As a result, almost as soon as the Ida Glenn No. 1 was completed, Galbreath was forced to begin work on several storage tanks to hold the production. The storage situation was eased significantly when the Prairie Oil and Gas Company, a part of the Standard Oil Company complex, connected its 6-inch pipeline running between its refinery at Whiting, Indiana, and the Red Fork Field to the discovery at Glenn Pool with a 2-inch line on February 16, 1906. The outlet was so important that with the connection of the last string of pipe, local oilmen christened the line with a bottle of wine, despite the fact that liquor was prohibited in Indian Territory.[41]

Until the summer of 1906, many people had refered to Galbreath's discovery simply as the Glenn wells in deference to the landowners. However the prominent citizens of several nearby towns—Mounds to the southwest, Tulsa to the northeast, and Sapulpa to the northwest—realized the potential publicity to be gained if their communities were associated with the discovery and called the strike the Mounds Oil Field, the Sapulpa Pool, or the Tulsa Field. Galbreath, however, prefered to name the pool in honor of the landowners on which the Ida Glenn was drilled. And after a meeting in the old Reeder Building in Tulsa, he succeeded in persuading his partners officially to name the field Glenn Pool. Later, as additional discoveries were made near the original field, the name Glennpool Field was given to the conglomeration of strikes.[42]

By Christmas of 1905, it was obvious to Galbreath that he had located a huge deposit of oil. When he returned to his home in Red Fork for the holidays, he began making plans to move his family to Tulsa, already becoming the center for many of the petroleum investors and companies operating in Indian Territory. So rapid was Glenn Pool's development that by the early summer of 1906 three producing wells a day were being completed in the field.[43]

Because of Glenn Pool's tremendous outpouring, by 1906 Oklahoma had outstripped the other states in the Mid-Continent-Gulf Coast Region (Texas, Kansas, Arkansas, and Louisiana) in crude production. That year Oklahoma produced 18,091,000 barrels of oil. At the same time, Oklahoma moved from 8th among the oil producing states in 1905, to second, behind only California, in 1906. The following year, 1907, Oklahoma's annual output of 43,524,000 barrels of oil made it number one among the oil-producing states.[44]

CHAPTER SEVEN

FLUSH PRODUCTION

The strike at Glenn Pool proved to be a tremendous boon to early investors. Of the field's more than 1,100 wells, only 2 percent of them were dry holes, and the field quickly grew from its original 80 to more than 8,000 acres. Such a drilling boom brought a tremendous amount of wealth to the region. By the end of 1907, approximately 95 companies operating in the pool had already spent $4,548,000 in drilling, $5,000,000 for lease purchases, $3,000,000 in salaries, and $15,323,125 for timber and steel for construction. In 1908, their return for the investment was 20,494,313 barrels of crude. Although production later declined, as late as 1935 the region had an output of 1,904,000 barrels annually.[1]

One of the first to arrive at Glenn Pool was Harry F. Sinclair. Although born in West Virginia, Sinclair was raised in Kansas, where he initially had been trained as a druggist. His first venture into the oil business was the sale of lumber for local oil well construction. As his contacts with drillers grew, Sinclair began purchasing small leases and forming companies to develop his holdings.[2]

In 1904, a syndicate headed by Sinclair enjoyed extraordinary success in some wildcats in southwest Oklahoma near Kiowa, which produced a $100,000 profit. With the Kiowa money, Sinclair developed the Canary Field, in present-day Washington County, in which he eventually drilled 100 wells. When Sinclair heard of Galbreath's discovery, he promptly headed for Glenn Pool, where, as he later recalled, within a short time he became "a millionaire."[3]

Throughout his early Oklahoma activity, Sinclair had maintained his home in Kansas, where, by 1907, he was regarded as the "richest man" in the state. His total holdings in 1913 included 62 oil companies and 8 drilling rigs. In addition, he was a partner with his brother Earl W. Sinclair in a Tulsa bank. By 1916, he was the largest independent oilman in the Mid-Continent Region, and on May 1, 1916, he pooled his holdings to form the Sinclair Oil and Refining Company. At the time, the *National Petroleum News* hailed Sinclair's $50 million action as the "Biggest Independent Oil Deal [ever] Completed. Within a month it was the ninth largest oil company in America.[4]

The Getty Oil Company, through its subsidiary Tidewater Oil Company, was another early developer of Glenn Pool. Tidewater, which Getty acquired in 1952, had been involved in Glenn Pool, through its subsidiaries Associated Producers and the Oklahoma Oil Company, almost from the beginning. Shortly after Glenn Pool's opening, David P. Connelly, Tidewater's Oklahoma superintendent, purchased 160 acres near the Ida Glenn No. 1 for $17,500. Ironically, when Tidewater's corporate offices first learned of the transaction, Connelly was criticized severely for such a rash purchase. However, the initial Associated Oil Company well on the property was completed as a 1,500-barrel-per-day producer. Within two years, Connelly's lease had produced 1,100,000 barrels of crude.[5]

By March, 1907, the Tidewater organization had 9 producing wells in Glenn Pool, and within another three months these wells were producing so much crude that the company constructed three nearby 35,000-barrel steel storage tanks to handle the output. By the end of the first year of Glenn Pool's life, the Associated Oil Company had completed 14 producing wells in the field, with a total output of 9,000 barrels daily, and had another well under way. Because of their aggressive drilling policy, by 1906 Tidewater, with approximately 20 percent of Glenn Pool's production, was the field's second major producer, falling behind only Galbreath and Company.[6]

There was an early attempt at Glenn Pool to curb excessive drilling. By common consent, all wells in the pool were supposed to be drilled near section lines, with a mininum of 150 feet between the well site and the road. In addition, only five wells were to be sunk to the half-mile. However, because of the huge influx of numerous oil companies these restrictions virtually were ignored in the mad scramble.[7]

As W. L. Connelly, who eventually became Vice-President and Chairman of the Board of the Sin-

A Glenn Pool gusher. Such production quickly touched off a rush to the area. (Sapulpa Historical Museum).

clair Oil and Gas Company, recalled, "there was no clearly defined well-spacing in that era [during the Glenn Pool boom], except insofar as an operator or company chose to utilize the principle of putting wells far enough apart to get maximum recovery from the oil sands. It was not uncommon for wells to be drilled on seven- or eight-acre spacing." As Connelly explained, the wooden derricks "could be constructed, and the cable tool drilling tools got into operation, in five or six days," and the drilling of a 1,400 or 2,000 foot well "required only another eighteen to twenty days." As a result "there were as many as fifty to one hundred wells drilling at a given time, and, of course, hundreds of earlier wells."[8]

Connelly recalled that "the whole countryside had a bristling appearance from the hundreds of wooden derricks rushing into the sky." Construction of the derricks was a specialized task. According to David Anderson, a long-time area oilman, the walking beams, oscillating beams used to transmit the reciprocating vertical motion of the steam boiler to the drilling tools, were 14 feet long. They were supported by the sampson post, which was 14 feet tall and 18 inches square. The band wheel, which transmitted the belt-driven power of the boiler to the main crank shaft, was supported by a shaft that was 13.5 feet long and which could hold between 1,000 and 2,000 feet of cable. Because they were constructed of wood, spliced with carriage bolts and iron straps, the shackle rods, which extended from the pump house to the wheels, often broke during drilling operations.[9]

At least twice during Glenn Pool's boom years, the field was swept by severe wind storms. On both occasions the high winds destroyed most of the field's wooden derricks. After the first storm, local oilmen rebuilt the wooden structures, but after the second storm the wood was replaced by steel.[10]

Making the scene even more "eerie at night" were the "scores of visible gas flares" disposing of unwanted natural gas. Although the operators used "as much of the gas produced...as...[they] might need for fuel purposes on the location,...the rest was flared." "It is easy to look back and call this wastefulness," Connelly commented, but this was before the "advent of gasoline plants, which were to convert natural gas into motor fuels, and, because Glenn Pool was developed so fast, "there were really no marketing facilities for the natural gas."[11]

Many of Glenn Pool's wells were completed with an initial flow of between 1 million and 30 million cubic feet of natural gas daily. Because, as Connelly explained, the natural gas was worthless except as fuel for the drilling rigs, many oilmen allowed the gas to spew into the air, believing that somehow by allowing the gas to continue to flow the wells would begin producing oil. A few wells blew gas for a long as eight weeks, and it was estimated that by 1912, a total of 50 billion cubic feet of natural gas had been vented into the atmosphere at Glenn Pool. As a result, within three years of the beginning of development, there was not sufficient gas pressure remaining in the field for the wells to flow naturally.[12]

Adding to the waste at Glenn Pool was the prevailing "Law of Capture," which gave the crude to whichever company pumped it to the surface, even if the crude was being pulled from under adjoining leases. Because of such a philosophy, the only way an oilman could prevent his leases from being drained by wells on neighboring leases was to drill as close to the property line as possible.

As described by W. L. Connelly, who eventually became Vice-President and Chairman of the Board of the Sinclair Oil and Gas Company, "the whole countryside [around the town of Glenpool] had a bristling appearance from the hundreds of wooden derricks rushing into the sky." In the right foreground a 55,000-barrel storage tank is under construction. (University of Tulsa, McFarlin Library).

Such action resulted in dramatic line fights as neighboring leaseholders drilled well after well along property lines opposite one another.[13]

One of the classic Glenn Pool line fights developed between Robert Galbreath and David Connelly, who held two adjoining 80-acre tracts. According to Frank Galbreath, Robert Galbreath's son, both men started a "line of wells 300 feet apart—the full half mile length of their eighty acres," with "seven or eight wells flowing into the biggest lake of oil I ever saw." All of the wells were "flowing from two hundred fifty to fifty-five hundred barrels each and every day," and "the greater part of the oil was wasted," the younger Galbreath admitted.[14]

There were several productive horizons in the field: the Big Lime at 695 feet; the Oswego at 815 feet; the Perryman, at 1,003 feet; the Red Fork at 1,225 feet; the Glenn at 1,350 feet; the Taneha at 1,550 feet; the Burgess at 1,710 feet; the Mississippi at 1,875 feet; the Sylvan at 2,100 feet; the Wilcox, which varried in depth, but was somewhere between 2,100 feet and 2,160 feet; the Hominy at 2,160 feet; and the Turkey Mountain at 2,250 feet. There also were reports of the Mounds Sand at 2,390 feet, and another unnamed productive horizon at the 2,530 foot level. However, most of the production came from five horizons: the Glenn Sand, which was approximately 20 feet thick and which averaged an initial daily production of between 40 and 600 barrels of crude; the Taneha formation, which was about 100 feet thick, and had an initial daily production of between 10 and 100 barrels of oil; the Wilcox Sand, which was only about 10 feet thick, but which had an initial production of between 70 and 300 barrels of oil daily; the Hominy horizon, about 30 feet thick, which initially produced between 25 and 100 barrels of oil daily; and the Turkey Mountain formation, which, with a thickness of approximately 20 feet, flowed between 30 and 2,000 barrels daily.[15]

By the summer of 1906, Glenn Pool's total production equaled more than half of the entire territory's output, with no signs of development slowing. Galbreath and Company had already

Note how close the two wells in the foreground of this photograph have been drilled. The early development at Glenn Pool was governed by the "Law of Capture;" that is, whoever brought the oil to the surface owned it, no matter from under which lease it came. Thus, most of the field was overdrilled, with line fights common, as competing oil companies sought to protect their investments. (Beryl D. Ford).

completed 21 wells, flowing 14,000 barrels daily, with 2 other wells drilling, and sites being prepared for 5 more. The Associated Producers Company reported 14 producers, flowing 4,600 barrels of oil, 1 gasser, and 1 well making hole. The Creek Oil Company was producing 11,000 barrels daily from 11 producers. Milliken, on the Gilcrease lease alone, had 8 producers with an output of 5,600 barrels per day. Seven other companies—E. J. Bonacker; Quaker Oil Company; Prairie Oil and Gas; Shawnee Oil Company; Litchfield, Sawyer and Company; Laurel Oil Company; and Selby Oil Company reported a total of 33 producers flowing 3,000, 2,400, 2,400, 1,800, 1,500, 1,400, and 1,400 barrels respectively.[16]

Ironically, the high production proved to be the curse of the field. As early as 1906, the Prairie Oil and Gas Company had completed a pipeline into Glenn Pool. In addition, the St. Louis-San Francisco Railroad had built a 1,000-foot loading rack at Kiefer, just to the southeast of Glenn Pool, that could accommodate 25 railroad tank cars at a time. Likewise the Midland Valley Railroad opened a loading rack at Jenks, between Tulsa and Glenn Pool. Also the Gulf Pipe Line Company built approximately 30 miles of gathering lines inside the field, as well as ten 55,000 barrels iron storage tanks. As early as February, 1907, Gulf began shiping crude by rail to its refinery in Port Arthur, Texas.[17]

Construction of another line, called the Battle Axe Railroad, was planned to penetrate the heart of the field. However, the promoter of the project was jailed for utilizing county funds to finance the project, and work was halted. All was not lost to area oilmen, however, because prior to the venture's failing quite a bit of the right-of-way had already been graded and the raised portions were used as dams to hold water for drilling activity.[18]

Despite these efforts, nevertheless, the vast quantities of petroleum being produced caused a

The typical wooden storage tanks used at Glenn Pool during its early development are shown in this photograph. The main drawback of the wooden tanks was their size. They simply could not be built sufficiently large to hold the tremendous production without experiencing serious problems with leakage. (Oklahoma Historical Society).

Because of the inadequate storage facilities of Glenn Pool, many oilmen were forced to pump their oil into earthen pits such as this. Completely unsuitable for long term storage, such pits were notorious for their high seepage and evaporation rates. (Beryl D. Ford).

Polecat Creek ran from near Sapulpa, through Glenn Pool, and into the Arkansas River. Because it was located close to many producing wells it was not uncommon for the water to be covered with crude from leaking storage tanks, earthen pits, and pipelines. Often the crude was set on fire, either accidentally or by farmers wanting to keep the oil away from their farmland or livestock. In this photograph, taken on February 12, 1908, the entire length of the Polecat Creek is burning. (Sapulpa Historical Museum).

A view of Polecat Creek after the February, 1908, fire. Note the destruction of undergrowth and trees along the banks of the creek. (Sapulpa Historical Museum).

severe storage problem as production exceeded transportation facilities. To store the excess production, oilmen frantically erected both wooden and metal storage tanks. "A large force of men and teams" already was busy by January, 1907, "making excavations in the earth with plows and scrapers to hold the oil that cannot now be received by the pipe lines for lack of capacity." These proved to be stopgap measures, and by January, 1907, it was estimated that 5,000 barrels of oil daily were lost through fire, evaporation, or seepage.[19]

Few of the oil storage containers, either wooden, steel, or earth, were completely leakproof. Much of the crude escaped to rush "down streams like a flash flood, before finally coming to rest in [a] natural surface pool." It was not uncommon for the crude that had escaped into nearby Polecat Creek to catch on fire for miles and fill the air with black smoke.[20]

From the air, such lakes of oil looked like any other waterway, and waterfowl often landed on the crude. However, once covered with oil it was impossible for them to take off again, and with their "oil smeared wings helplessly flopping," they were condemned to die. Quick to take advantage of the fowls' plight, "restaurants would rush to the scene with empty buggies to load" with waterfowl. As one Glenn Pool observer recalled, in many area cafes "roast duck was cheap." Ducks were not the only victims. It was reported that three cows died, "their mouths full of oil."[21]

Such a huge outpouring of crude also drastically affected the price. In February, 1907, Galbreath was receiving 52¢ a barrel for his crude, but by July of that year the price had dropped to 42¢. Vast amounts of Glenn Pool oil were sold at 25¢ a barrel, and throughout its boom period the average price in the field was 30¢ a barrel. With the average well cost pegged at $5,000, it took a lot of 30¢ oil to pay for the drilling operation. While many oilmen realized the foolishness of continued drilling in the face of low prices, local landowners refused to cooperate in any effort to ease the oil glut. "Better some oil going out and a few dollars coming in than keeping the wells shut down," they rationalized.[22]

The crude itself was a very high grade and should have commanded a higher price. Black or dark green, Glenn Pool's crude rated 28-32 degrees gravity, with a paraffin base. Its asphalt content was low, generally less than 50 percent.[23]

Construction of a nearby refinery obviously was one solution to the oversupply of oil. As early as February, 1906, H. H. Tucker, Secretary for the Uncle Sam Refinery Company in Cherryvale, Kansas, asked the Tulsa Commercial Club to finance a 15-acre refinery site on the west side of the Arkansas River near the Frisco Railroad bridge. In an effort to convince local businessmen to underwrite the cost of acquiring the property, Tucker offered to sign a $10,000 bond as insurance that a 600 barrel-per-day refinery would be in operation by July 1, 1906.[24]

With 15 employees, such a refinery complex would be a great boon to Tulsa and the Commercial Club appointed William Stryker, D. C. Rose, and J. E. Piersol as a committee to examine the property. In addition they were authorized to raise money for the project. Eventually, the NW 1/4, NW 1/4, SW 1/4 of Section 11-T19N-R12E was purchased. Although the Commercial Club fulfilled its part of the agreement, the refinery never was completed when it became obvious that several pipelines then under construction would link Glenn Pool to established refinery complexes in Texas, which were capable of handling far more than the proposed 600-barrel per-day operation.[25]

As early as December, 1906, Galbreath had announced plans for a $20 million investment by British financiers to form the Mid-Continent-Port Arthur Pipe Line and Refining Company. Galbreath's promised capital did not materialize and the venture faded. Even so, the fact that Glenn Pool's crude was far superior to any other available at that time meant that other companies were ready to invest the necessary money to reach the oil.[26]

It was a simple matter of economics. Glenn Pool's crude made such excellent gasoline and kerosine that refining companies could afford to pay as much as 78¢ a barrel to transport it by rail to Port Arthur, Texas, and still make a profit. As a result, in the summer of 1907, 860 railroad tank cars were filled with Glenn Pool crude at Jenks, and another 555 tank cars were loaded at Kiefer to rumble "southward day and night." Construction of a pipeline linking the Texas Gulf Coast to Glenn Pool obviously would reduce the transportation cost significantly, but in 1906 such a project was considered "audacious" by many oilmen.[27]

Nonetheless, in September of that year, officials of The Texas Company decided that the

Fire was another serious result of flush production at Glenn Pool. With inadequate storage facilities and the use of earthen storage tanks, which easily caught on fire, the danger of a disaster increased. (Sapulpa Historical Museum).

To offset the chance of burning oil spreading to nearby tanks, the large storage tanks were surrounded by earthen dikes to catch any burning oil. In this photograph, six teams of scrapers are preparing such a dike. Note that the third driver from the left is a woman. (Sapulpa Historical Museum).

company would extend its Port Arthur to Humble, Texas pipeline northward to Tulsa. Although the estimated cost of the project was between $4 million and $10 million, within 30 days of the decision surveyors were staking the route which ran from The Texas Company's tank farm, about two miles south of Tulsa on the Clinton allotment, southward to Okmulgee and Henryetta, before it angled slightly west to Atoka and Colgate. The line crossed the Red River to the east of the MKT's (Missouri, Kansas, and Texas Railroad) bridge, and then ran south to Dallas, Texas. There it turned east toward Humble, where the line joined an already existing pipeline into Port Arthur.[28]

Work crews started building southward from Tulsa on Feburary 4, 1907, and the 473 miles of 8-inch line was completed in a little more than a year. The first Glenn Pool crude reached the Port Arthur refinery in November, 1907. Although the project cost $5,641,000, "the line was a moneymaker from the start."[29]

At about the same time as The Texas Company's decision, the Gulf Production Company, in the fall of 1906, began planning another 8-inch pipeline that would stretch 413 miles from Glenn Pool to Sour Lake, Texas, which at that time was the northernmost limit of the firm's pipeline system. Because the line would have to cross both the Red and Canadian rivers, Gulf Refining Company officials reached an agreement with the Frisco Railroad to allow its construction crews to utilize existing railroad trestles over the rivers. The projected cost, including gathering lines, pump stations, and other equipment, was $3,000 per mile.[30]

To oversee the venture, the Gulf Pipe Line Company of Texas was created, and as soon as W. L. Mellon, one of Gulf's principal owners, received permission from the Department of the Interior to construct the pipeline across Indian Territory, work got under way. In preparation for the necessary federal approval, while Mellon was in Washington, D. C., H. Y. Arnold had been sent to Tulsa to secure sites for tankage, pump stations, etc. Eventually 55 railroad cars of pipe were unloaded at Tulsa for the project. At the same time, other Gulf representatives began purchasing right-of-way.[31]

Two crews, totaling 3,000 men, were hired. Half were sent to Texas to work north and half were sent to Tulsa to build southward. Work began at Sour Lake on February 3, 1907, and at Tulsa on March 12 of that same year.[32]

In Texas the line ran from Sour Lake north across Hardin, Tyler, Angelina, Nacogdoches, Rusk, Smith, Wood, Franklin, Hopkins, Delta, and Lamar counties, before crossing into Indian Territory just west of the Frisco Railroad bridge. Inside Indian Territory, the route passed a little to the west of South McAlester, then north to Okmulgee, and on to Glenn Pool, generally parallel to the 96th Meridian. It was completed in August, 1907.[33]

When finished the Gulf pipeline held the record for speed in pipeline construction, and was the longest piece of pipeline laid in the United States by any company other than Standard Oil. Along its right-of-way, six pumping stations, at intervals of approximately 80 miles, were built to push the oil through the line. Two of the stations were located at Watkins and Chambers in Indian Territory. The other four were built at Lenior, Big Sandy, Lufkin, and Sour Lake in Texas.[34]

Along with construction of the pipeline, Gulf Oil Company also decided to become more involved in Glenn Pool's production. To oversee the production phase of its operation, Gulf sent Frank A. Leovy to Tulsa with instructions to purchase already producing leases. Leovy, a devoted sailor, named the new company the Gypsy Oil Company, after his favorite boat the *Gypsy*.[35]

Like The Texas Company's line, Gulf's pipeline proved to be very profitable, and by 1908, 5/7th of Gulf's total production came from Oklahoma wells. Because Gulf's and The Texas Company's Gulf Coast refineries had the capability of handling 7 million barrels of Glenn Pool oil annually, it was thought that the completion of the two pipelines would eliminate the glut of oil. However, by 1909 Gulf's production had grown to the extent that it was necessary to increase its Port Arthur-Glenn Pool pipeline capacity from 16,000 to 21,000 barrels daily.[36]

In addition to the Gulf and Texas Company pipelines, Harry Sinclair tied Glenn Pool to a 478-mile 8-inch pipeline system that terminated at the Coffeville, Kansas, refinery. Eventually, in 1918, Sinclair's system was expanded northeast to East Chicago, Illinois. This gave Glenn Pool oilmen an outlet to the Great Lakes.[37]

After the oil pipelines, several natural gas lines were built into Glenn Pool in an effort to tap the

As shown in this panorama, the huge 55,000 barrel tanks eventually dotted the Glenn Pool landscape. (University of Tulsa, McFarlin Library).

The St. Louis and San Francisco Railroad depot in Kiefer. The railroad offered an outlet for Glenn Pool's crude prior to the completion of pipelines into the region. (Sapulpa Historical Museum).

The St. Louis and San Francisco Railroad's 1,000 foot loading rack at Kiefer. It could handle 25 railroad tank cars at one time. (Sapulpa Historical Museum).

field's huge output of natural gas. Shortly after the completion of the Ida Glenn No. 1, Galbreath and his partners had formed the Galbreath Gas Company and constructed a 6-inch gas line across the Arkansas River to Tulsa. However, the project was abandoned after Galbreath got into a price war with the Osage and Oklahoma Gas Company and the price of natural gas plunged to 5¢ per thousand cubic feet.[38]

In 1906, the Oklahoma Natural Gas Company had been formed with the idea of connecting the gas wells in the Cleveland Field with Oklahoma City. However, before the pipeline could be built, Cleveland's production had dropped to the point that it could no longer supply the demand. As a result, in the fall of 1906, after the Osage and Oklahoma Gas Company had gained control of the Oklahoma Natural Gas Company, it decided to tap Glenn Pool's gas output.[39]

Construction of a 16-inch natural gas pipeline between Tulsa County and Oklahoma City was started in the spring of 1907. Eventually, it was connected to two natural gas wells near Sapulpa, about 5 miles northwest of Glenn Pool. Gas shipments began on December 18, 1907, and shortly afterward 8-inch branch lines soon were built to tie Guthrie and Shawnee into the system.[40]

In spite of the lines connecting Glenn Pool with the Gulf Coast, Great Lakes, and Oklahoma City, Glenn Pool's prolific output of oil and natural gas still exceeded all available markets and storage facilities. Eventually, the huge waste, combined with overproduction and declining prices, forced Glenn Pool oilmen into another attempt to regulate production. Although the effort was somewhat a success among already producing wells, newly completed gushers prevented any long-term planning.[41]

Contributing to the glut of oil was the discovery of the Taneha Pool, a satellite of Glenn Pool, in February, 1909. Rail shipments of oil from Taneha began in July, 1909, and although it was drilled more slowly than Glenn Pool, by November, 1909, Taneha's daily production had reached 40,000 barrels of crude daily. Peak monthly production at Glenn Pool was reached in March, 1911, when the field's output reached 2,584,464 barrels of oil per month.[42]

In 1907, during the heyday of Glenn Pool's production, Oklahoma and Indian territories were joined into the state of Oklahoma. From the beginning of the Oklahoma petroleum industry until statehood, there had been three distinct governmental bodies regulating the production of crude—the federal government, officials of Oklahoma Territory, and various tribal governments. Of these three, federal and tribal officials were more concerned with the payment of royalties to Indians instead of providing guidelines for the creation of a stable petroleum industry. However, as the oil industry grew, tribal officials had enacted some general regulatory laws and a tax of 10¢ per 40-gallon barrel of crude. Even so, tribal rules did not hinder development as much as federal regulations, which were codified in June, 1898 in "An Act for the Protection of the People of the Indian Territory, and For other Purposes."[43]

Generally called the Curtis Act, this law gave the Secretary of the Interior the final say in the leasing of tribal land. Disturbed by the erosion of their regulatory authority, the Cherokee Nation had appealed the Curtis Act to the United States Supreme Court, before losing. As a result, in the time between 1898 and statehood in 1907, most drilling activity in Indian Territory was very restricted, with federal regulations—most of which resulted from bureaucratic squabbling—actually encouraging waste in many instances. For example, one rule called for the drilling of one well

The Sapulpa Refining Company operated a refining complex in Sapulpa, just to the northwest of Glenpool, which refined much of Glenn Pool's crude. By 1909, the facility was in full operation. (Sapulpa Historical Museum).

annually on every lease. Obviously such a regulation did little to encourage the orderly development of a new strike. Instead, it contributed to the mad rush to sink more and more wells, whether they were necessary or not. While failing to stem the extravagant waste at Glenn Pool, such regulations did entangle the oilmen in a maze of rules.[44]

During this same time span, in the western half of the new state, the government of Oklahoma Territory had enacted a few regulatory tatutes pertaining to the petroleum industry. Most of its legislation was constructive, however, as the territorial legislators attempted to prevent a repeat of the notorious waste of oil and natural gas that had occured in the Indian Territory. As early as 1895 the office of Oil Inspector had been established, but, because no strike in Oklahoma Territory compared to what was being found in Indian Territory, there was little for him to do.[45]

With statehood came the end of much of the federal regulation over Indian Territory, as state officials assumed the task. Frustrated in their attempts to deal with federal officials, most Tulsa and Glenn Pool area oilmen welcomed statehood as an end to their problems. In fact, Frank Galbreath later recalled that when word was received that President Theodore Roosevelt had signed Oklahoma's statehood proclamation "the town went wild." "Crowds thronged the streets from daylight far into the night, yelling, singing and 'whooping it up,' " with "some of the tougher element" growing "even more vociferous, and...[giving] vent to their high spirits by shooting revolvers, shotguns, sixshooters, any thing to make a noise." "Whistles blew, bells rang and parades were held" to mark the occasion, and "kegs of whiskey were placed on the counters of several of the oil well supply stores where oilmen congregated, and each visitor helped himself from a tin cup."[46]

In anticipation of statehood and the accompanying relaxation of federal restrictions, several companies interested in acquiring Glenn Pool's

Fire at the Sapulpa Refining Company tank farm in 1911. The tanks on the right are engulfed in flames, while firemen pour water on the tank to the left in hopes of keeping its temperature sufficiently low so it will not boil over and catch fire. (Sapulpa Historical Museum).

high-gravity oil already had completed pipelines to the borders of Oklahoma. There work had stopped while the companies awaited the end of federal control. Faced with the prospect of the companies beginning work across the state's boundaries before the legislature had time to formulate rules governing pipelines, Governor Charles N. Haskell ordered the National Guard to seal the state's boundaries to pipeline construction. This was the first time martial law was used to control Oklahoma's petroleum industry.[47]

While the Oklahoma constitution virtually ignored the regulation of oil wells, pipelines were given close scrutiny, and one the first acts of the newly elected state legislature was to require that all pipeline companies within the state be chartered. The solons also provided for compensation to be paid property owners for damage caused by pipeline constructon, limited gas line pressure to 300 pounds per square inch, and banned the shipment of natural gas beyond the state's boundaries. Supervision of pipeline work was relegated to the state's Chief Mining Inspector. Control over the oil business was given to the Oklahoma Corporation Commission by the state constitution.[48]

One source of confusion resulting from the multitude of regulatory agencies prior to statehood was the question of Indian leases in state courts. During the period of federal control, federal courts had handled the question of Indian leases. With statehood jurisdiction should have been transferred to state courts; however, there was no corresponding state law to give the state judges that power. As a result, as late as 1909 federal agents were maintained in Oklahoma to insure the fair treatment of Indian minors.[49]

Recognized as one of the major oil producing area in America, Glenn Pool's discovery pushed Oklahoma into first place among America's oil producing states in 1907, with a cumulative output that year of 43,524,000 barrels. Glenn Pool's output peaked the following year, 1908, at

20,494,313 barrels of oil. Glenn Pool's tremendous outpouring of crude came at a time when the United States and the world were turning from coal to petroleum products as their primary source of energy. It was estimated that the United States Navy alone would need 2 million barrels of fuel oil every year once it completed its conversion from coal to oil-powered ships. In addition, with the beginning of the "age of the automobile" the demand for gasoline greatly increased. Fortunately, Glenn Pool's high-gravity was available to help meet the rising demand.[50]

CHAPTER EIGHT

GLENN POOL'S BOOM TOWNS

The strike at Glenn Pool and the accompanying scramble for riches led to the birth of several nearby oil boom towns. Four were especially prominent: Sapulpa, to the northwest; Kiefer, to the west; Mounds, to the southwest; and Jenks to the northeast. Sapulpa, Mounds, and Jenks all had been established prior to the the Glenn Pool discovery but grew phenomenally after the completion of the Ida Glenn No. 1. Kiefer, however, owed its birth to the nearby wells and grew almost overnight from a few buildings along a section line road to a bustling oil boom town. Several other communites were established in the oil fields around Kiefer, such as Taneha, which later became Bowden, and Sabo; however, the latter disappeared with the end of the boom.[1]

There was another community in the field, the town of Glenpool. However, while Glenpool was the home of many oil field workers, its leaders boasted of not having "a saloon in the town." Because of such an attitude, it missed much of the vice and violence that often accompanied a nearby oil strike.[2]

The town of Glenpool was started in the autumn of 1906 when the Midland Valley Railroad built a spur line from its main line at Jenks through the Glenn Pool oil field. The tracks ended near the home of Ida Glenn, and eventually a small depot, named Glen Pool Station, was constructed to handle the local rail traffic. Two years later, in 1908, the community had grown so large that it was given a post office. It too was named Glenpool. Although the community boomed during the heyday of the Glenn Pool strike, once the oil disappeared so did most of the town's inhabitants. Afterward the settlement rapidly declined.[3]

The oldest and largest of Glenn Pool's nearby boom towns was Sapulpa, which had been granted a post office on July 1, 1889. The community was named after a local inhabitant, James Sapulpa, near whose home the St. Louis and San Francisco Railroad had built a camp while constructing its tracks through the region. Inhabited largely by Indians, Sapulpa was an unpretentious rural agricultural settlement, with a population of 891 in 1900, economically dependent on local farming and ranching operations.[4]

Prior to statehood, Sapulpa had served as a Record Town for Recording District 8 of Indian Territory. After 1907 it became the county seat of Creek County. Because it was the largest community near the strike, Sapulpa became a mecca for many oilmen, and by 1910 the town's population had grown to 8,283.[5]

Only about four miles northwest of the Glenn Pool oil field, Sapulpa capitalized on its location, and for a while rivaled Tulsa as the oil center of the region. In fact the local Commercial Club, the forerunner of the Chamber of Commerce, erected a huge electric sign proclaiming "Sapulpa, the Oil City of the Southwest."[6]

After the completion of the Ida Glenn No. 1, "men came by the hundreds" from as far away as West Virginia and Pennsylvania "to look for rooms, homes, and office space in Sapulpa," Ted Waibel, a long-time inhabitant, recalled. With the pounding of screws and the clanking of bits making a continuous roar, "men with hats angled slightly over one eye and cigars tilted in the corner of their mouths flew past in buckboards," as they huried to the latest strike hoping to lease neighboring land."[7]

With them came "wagons so heavily loaded with casing-pipe that they cut deep into the sand and mud" roads. Usually only the teamster was available to load equipment on their wagons, and they became experts at manhandling the heavy oil field boilers. Turning their wagon "up on its side," the teamster first "put chains around the boiler," then hitched his team to it and rolled it onto the wagon. Then he would "'boom' or chain it to the wagon, [and] then with the same team would pull the wagon up on its wheels, and then he was ready to go."[8]

With the influx of thousands of oil field workers, "all available hotels and rooms" rapidly were filled in Sapulpa and the other nearby boom towns. The shortage of space became so critical that men slept in "tents, shacks, wagons or almost anything

The St. Louis and San Francisco Railroad Depot in Sapulpa. Because it was a well-established community prior to the Glenn Pool strike, Sapulpa was the jumping off place for many oilmen heading for the wells. Note the hacks to the right of the photographs waiting to carry passengers to the drilling area east and south of the town. The "dandy" in the center of the photograph has two grayhounds to accompany him on his ride. (Oklahoma Historical Society).

that offered shelter."At the height of the oil boom, 30 trains a day ran between Sapulpa and Tulsa.[9]

To handle the tremendous increase in traffic, the St. Louis and San Francisco Railroad constructed a modern new depot and yard facilities at Sapulpa. "One of the handsomest and best appointed passenger depots on the lines of the Frisco" when it opened, the rail yard and shop facilities were agumented with additional "acres of new yards and the laying of a score of miles of switch track." A four-stall machine shop, a blacksmith shop, a mill shop, two cinder pits, a stand-pipe, and a round house completed the complex. By 1908, Sapulpa was the terminus of four passenger and freight divisions of the Frisco, with the rail yards containing 40 miles of track and a payroll of nearly $125,000 every month.[10]

The Sapulpa and Interurban Railway connected Sapulpa with Glenn Pool. Many oil field workers would ride the train to Sapulpa and then catch the interurban to the nearby wells. Some, however, could not afford the interuban's fare and had to walk. For example, L. L. Fosburgh, who moved from Ohio to Sapulpa in 1906, had only 10¢ when he arrived. Tired and hungry, he purchased a dinner of cheese and crackers with the dime, and then started for the nearby wells on foot.[11]

By 1911 Sapulpa was a bustling city which boasted five banks—the American National Bank; the Sapulpa State Bank; the Farmers and Merchants Bank; the Oklahoma State Bank, which had previously been known as the Creek Bank & Trust; and the First National Bank. Two of the institutions closed, however, as the oil boom drew to a close. The Farmers and Merchants Bank ceased operations in 1912, and the Oklahoma State Bank in 1913. In addition, there was the modern Saint James Hotel, a pressed brick company, a steel oil storage tank factory; a sucker rod factory, an electric light and power plant, a cotton compress, two cotton gins, a water works, the Sapulpa Ice Company, and numerous other oil-field-related businesses.[12]

Chief among the community oil field supply establishments was the Boviard Supply Company. Incorporated in Kansas, the firm was a major supplier of oil field equipment in the early de-

An interior view of the St. Louis and San Francisco Railroad Depot in Sapulpa. The six clerks, two secretaries,and one messenger boy in this photograph were needed to handle the 30 trains daily that ran between Tulsa and Sapulpa during the height of the boom. (Oklahoma Historical Society).

velopment of the Mid-Continent Region. The firm's Sapulpa outlet was established in 1907 and was the first store ouside of Kansas. The initial facility quickly was expanded to also include a warehouse and repair shop for equipment.[13]

Sapulpa's most famous business attraction was the Harvey House Restaurant, said to provide the finest meals to rail travelers "between St. Louis and Fort Worth." As the trains neared Sapulpa, the passengers were asked how many planned to eat at the next stop. For six bits, 75¢, the Sapulpa's Harvey House offered "everything from *Consomme Printaniere Royale* to Broiled Plover on Toast."[14]

After the count was taken, word was telegraphed ahead to the Harvey House, and as soon as the train stopped, a "black gentleman in a white uniform bonging on a large gong to attract attention, appeared on the station platform to lead the customers to the restaurant. By the time they entered the door the "first course was being placed on the tables." A man had to be dressed in a coat and tie to be seated, and "no cursing...not even loud talk" was allowed in the dining room," which was set with such formality that the creases in the tablecloths had to be pointed toward the door. Customers were served by 33 "well starched and highly groomed waitresses," the Harvey Girls, "dressed in black poplin shirtwaists with stiff, white collars...the starched, white bib and apron...pinned in back."[15]

Such business endeavors notwithstanding, most of the community's sidewalks were unpaved, and people were considered lucky if they could make their way along a boardwalk instead of in the mud, which was knee deep in places. The dirt streets were so broken down by the heavy oil field traffic that after heavy rains "a team sometimes could

Constructing the interurban tracks through downtown Sapulpa. An extensive network of tracks connected most of the towns in Glenn Pool with one another. Many oil field workers lived in Sapulpa and worked in Glenn Pool, thereby avoiding much of the vice and violence of towns such as Kiefer. (Oklahoma Historical Society).

Sapulpa's St. James Hotel was the headquarters for many early oilmen in Glenn Pool. Note that while concrete sidewalks have been constructed, the street is still dirt. (Oklahoma Historical Society).

The Bovaird Supply Company's store in Sapulpa was the first of its kind in Oklahoma. This photograph was taken shortly after its opening in 1907. The third man from the right is W. J. Bovaird, the founder of the firm. (Sapulpa Historical Museum).

hardly pull through" the morass. It was not until 1909, two years after the electric street lights had been installed by the Sapulpa Electric Company, that the first street in the community was paved.[16]

Oklahoma still was very much of a frontier during the early development of Glenn Pool, and many Eastern oil brokers were taken aback by the "Wild West" atmosphere. W. L. Connelly, who rose to become Chairman of the Board of Sinclair Oil and Gas Company, had just started his association with the firm when Harry Sinclair sent him to Sapulpa to settle some lease questions in the pool. Connelly was not prepared for what he encountered.[17]

The problem centered around a lease to which Sinclair held the mineral rights, but a local farmer owned the surface rights. Like many farmers who did not own the mineral rights, this one, even though he was entitled to damages, would not allow the destruction of his crops by the oil workers as they prepared to drill, and demanded more damages than Sinclair thought were fair. Initially Howard Galbreath was sent to Glenn Pool to solve the problem, but he failed to reach an agreement with the farmer.[18]

After Galbreath's failure, Connelly, who at that time was in Sinclair's Independance, Kansas, office, was sent to Sapulpa to get things moving. Connelly decided on a direct confrontation, and ordered teams and wagons hauling rig timbers sent to the lease. When the procession started, Galbreath and Connelly followed in a buckboard. When the caravan reached the lease, Connelly recalled that "the farmer was sitting on a chair just inside the gate with a businesslike expression on his face and a still more businesslike rifle across his knees."[19]

As Connelly explained, "With the air of one in

Ben Hubbard, manager of the Bovaird Supply Company's Sapulpa store. (Sapulpa Historical Museum).

unquestionable command of the situation, I dropped from the buckboard and walked to the gate." "Keeping my eyes off the rifle as much as possible," he continued, "I delivered a dignified address." According to Connelly, "it was a masterful presentation of our case closing with a forceful demand that we be permitted to enter."[20]

When he finished talking, Connelly turned to the farmer who "eyed me with a stony composure." Shocked that the farmer's response, to his speech "was an unqualified 'No!'...to which he added, a concise appendix to the effect that he would shoot any white-livered so-and-so who tried to pass him," Connelly decided on more direct action. "With a gesture that meant I would stand for no nonsense, I told the teamster in the lead to drive in."[21]

"Not on your life!" the wagon driver replied. "Well, get down then," Connelly commanded. Quickly the teamster abandoned his seat, and Connelly took his place. Shouting "giddap," he started the wagon through the gate.[22]

"The horses barely had their heads past the gateposts," Connelly recalled, "when the farmer rasped a strident "Whoa!" After the horses had stopped, the farmer rose from his chair, "strolled over to the wagon," and jabbed "the muzzle of the rifle against my midriff." Growling "Now, you lousy thus-and-so, if you drive an inch farther I'll drill a hole through you," the farmer, according to Connelly, gave "the rifle a vicious poke" and "mumbled an insulting epithet which reflected upon my parentage." "I was profoundly impressed by his words and manner," Connelly later recalled, and by the time that Galbreath yelled for him to get off the wagon, "I was already down."[23]

The Dining Room of Sapulpa's Harvey House Restaurant. Five waitresses, two cooks, and a cashier are available to serve customers. Note the plates waiting to be used on the back of the counter. As quickly as a stool was vacated during the noon rush, the counter was cleaned, a new plate and silverware put out, and another customer seated. (Sapulpa Historical Museum).

Unpaved Dewey Avenue in Sapulpa during the Glenn Pool oil boom. Although the street was ankle deep in mud, paved sidewalks had been provided for pedestrians. Note the two teams of scrapers attempting to smooth the street. (Sapulpa Historical Museum).

Law enforcement officers raiding an illegal whiskey still in Glenn Pool. Although prohibition was enforced both before and after statehood, bootlegging was a profitable business during the oil boom. (Sapulpa Historical Museum).

Temporarly abandoning the effort, the two oilmen returned to Sapulpa and reported the farmer to the local Indian agent. After listening to the story, the agent sent Connelly and Galbreath back to the farm with an Indian policeman. Finally, under the protection of the officer, the well was spudded in.[24]

Sapulpa was legally dry throughout its history, both as part of Indian Territory and then as part of Oklahoma. As a result, "there was never a saloon legally operated" in the community. However, between the strike at Glenn Pool in 1905 and statehood in 1907, "suitcases of liquor" often were loaded aboard trains at Keystone, their last stop in "wet" Oklahoma Territory, before crossing the border into "dry" Indian Territory. Although no one accompanied the suitcase on the trip to Sapulpa, someone always was on hand to claim the baggage and then hurry on their way.[25]

Like other oil boom towns Sapulpa attracted a number of camp followers. While some pursued legitimate activities, others brought with them vice and violence. Several men were killed in shootings on the community's Main Street.[26]

At first Sapulpa's law enforcement facilities were woefully inadequate. The community's original jail was constructed of logs, and was too small to handle adequately the number of drunks arrested nightly, especially after a payday. When the jail was filled, local lawmen simply tied offenders to the logs around the outside of the structure.[27]

Although all of Glenn Pool's boom towns contained some unsavory characters, it was Kiefer, about four miles south of Sapulpa and eighteen miles from Tulsa, that acquired the worst reputation for violence and vice. Originally called Praper, the town's permanent residents consisted of "one or two" St. Louis and San Francisco Railroad

Originally called Praper, Kiefer started along the St. Louis and San Francisco Railroad tracks. By the time this photograph was taken, several shotgun houses, clapboard stores, and tents had been added to the original community. The two large buildings probably were hotels catering to the oil field workers. (Sapulpa Historical Museum).

By the time Praper's name was changed to Kiefer in 1906, the community was an oil boom town. Note the large number of boxcars bringing supplies and equipment to the nearby wells. Most homes in the photograph have their own outdoor privy. (Oklahoma Historical Society).

An interior view of one of Kiefer's general stores. Such establishments carried a myriad of supplies ranging from the canned goods, on the left, to the rack of automobile springs, on the right. Dominating the store is a large stove, and, although a cuspidor was provided for customers, the tobacco stains on the floor indicate that not all took advantge of it. (Sapulpa Historical Museum).

workers housed in a makeshift boxcar depot. In 1906 the community's name was changed to Kiefer, in honor of Smith Kiefer, a long-time local inhabitant. The town happened to have the closest railroad stop to the Glenn Pool strike, and by November, 1906, it was an established oil boom town.[28]

By then Kiefer was home to several hundred oilmen, workers, prostitutes, gamblers, bootleggers, and honest folks living in a "few" wooden frame buildings, stretching along the east and west sides of the railroad tracks. Most of the structures consisted of nothing but "a plain front built along the street, joining each other," with a tent in back of the front. By 1910 Kiefer's population was in excess of 2,000, and eventually it swelled to an estimated 8,000 to 10,000.[29]

Charlie Shobe, who moved to Kiefer from Ohio in 1907 to work for the Franchot Oil Company, recalled that "it seemed that confusion was everywhere" when he first arrived in the town. "Great masses of heavy wagons pulled by teams of ten to twenty mules or horses were going, or trying to go, in every direction....Steam boilers, rig timbers, pipe, wooden tanks, all sorts of oilfield stuff were lining the muddy streets," he continued. "The people, masses of them, surged back and forth, back and forth, on the dirt sidewalks and board porches fronting little stores which had been thrown up to cater to the needs of the thousands of people who had come here."[30]

Bob Gaberdiel had made the trip with Shobe, and the two decided to eat before trying to find the Franchot Oil Company office. "The little cafe was jam-packed with men," according to Shobe. "The counter was full, and men lined the walls waiting for a seat or just to get a sandwich handed to them which they could eat standing up....We joined the 'wall standers,' " he continued.[31]

Because it was late in the evening before they were able to eat, Shobe and Gaberdiel decided to look for a place to spend the night and start work the next morning. "We walked all over Kiefer ask-

A typical oil boom home built by Charlie Shobe and Bob Gaberdiel near Kiefer. The wooden frame structure was covered with canvas, to keep out the weather, and wooden boards, to keep out animals. Here Shobe and Gaberdiel pose with two of their friends and their dog on the front porch. (Sapulpa Historical Museum).

ing at every place we saw that had a sign reading 'Rooms,' "but, according to Shobe, each time they inquired they were told that the rooms were rented "by the week and even if a fellow leaves, he has a dozen buddies waiting to take his place." Finally, Shobe recalled, "we drifted over to a big tent stretched on a little piece of high ground with a big swampy slough surrounding it." "You had to walk across the stagnant slough on planks," he continued, "but you had nice dry ground once you were across."[32]

Luckily, the tent was the Franchot Oil Company headquarters. They were told that they would start work at 5:00 a.m. the next morning, and that they could lay some one-by-twelve inch boards on the ground and sleep on them that night. After borrowing some blankets and rolling up their coats for pillows, Shobe and Gaberdiel went to sleep.[33]

They were awakened early the next morning. There was no time to shave, and that baths were reserved for weekends. They were fed in a company "eating tent," given two cots in the "sleeping tent," and sent to work.[34]

Shobe was assigned to a pumping lease where the production of 30 wells was funneled into several wooden tanks. He had to watch the tanks, and whenever one was full then divert the flow into another tank. The weather was cold, and while "crossing a little gully" Shobe fell through the ice and suddenly found himself "plumb up to my knees in icy, muddy, oily slush." "Mud, ice and slush went into my mouth, nose, and ears," he explained, and after struggling to regain his footing he returned to camp and went to sleep, forgetting about the tanks for which he was responsible.[35]

"I slept soundly," Shobe recalled, "thinking the

Looking east on Kiefer's Main Street at the start of the Glenn Pool boom. The streets are unpaved, there are no sidewalks, and open ditches are the only drainage provided. However, the general store on the right has built a large floored area in front to display merchandise and placed rocks on the ground to allow customers to get out of the mud. Note the exposed pipeline beside the road. (Sapulpa Historical Museum).

boss would never know I had not made my rounds." Without his supervision the tanks quickly filled and overflowed. "Oil was everywhere" when he awoke.[36]

After their first payday Shobe and Gaberdiel ordered a tent from Montgomery Ward and borrowed a lease team to carry it to a "quiet little grove of trees near the lease." "We bought an old rusty iron bedstead and a straw tick," Shobe continued, and filled the tick with fresh hay. They used a "couple of flour sacks filled with hay" for pillows, and with a few "granite pots and pans for dishes, and a coal oil lamp" they set up housekeeping.[37]

"At first we cooked on a campfire with an old Dutch oven we'd bought at a second-hand store," Shobe explained. To weatherproof their tent, Shobe and Gaberdiel "boxed up" the structure with boards. Later, when they had acquired a "second-hand wood cookstove.... We cut a hole in the side of the tent and by using an elbow pipe we had," according to Shobe, "a pretty snug set-up."[38]

Kiefer received a major boost as the region's premier boom town in November, 1906, when the Gulf Pipe Line Company selected a site just north of the community for its crude oil loading racks. At about the same time, the railroad constructed a siding that could hold about 40 railcars near Kiefer's depot. Afterward, the town quickly became a major shipping point for Glenn Pool crude. By June, 1907, 154,000 barrels of oil were shipped monthly by the Prairie Oil and Gas Company and the Gulf Pipe Line Company from Kiefer's loading racks.[39]

The shipping facilities brought workers, and the workers attracted businesses that catered to their needs. Mercantile houses, tank shops, restaurants, and lumber yards hurriedly were built,

Kiefer's Main Street after the boom was well under way. Little has changed. The street still is not paved, there are few wooden sidewalks, and drainage still is provided by open ditches. On the right is the Texas Place, which offered cold drinks and rooms. Such establishments usually were fronts for saloons, gambling dens, and houses of prostitution. (Sapulpa Historical Museum).

An impromptu band of Glenn Pool oil field workers. While the men on the right are playing more common instruments, the two on the left are using pieces of pipe as a horn and as a percussion instrument. The bottles on the table are filled with water to produce different tones when blown into. (Sapulpa historical Museum).

Kiefer's Town Band posing on a raised platform from which they played. Note the signs on the clapboard building in the back. Horse Shoe Tobacco offered prizes for certain numbers of tags, and the bottom three posters advertise an uncoming fight featuring Carl Morris, "the great white hope." (Sapulpa Historical Museum).

The second man from the right is Carl Morris, a local resident who switched occupations from a railroad engineer to a professional prize fighter. Left to right are Max Meyer, Amos Midman, Morris, and Babe Banks. Morris fought several fights in an open arena between Sapulpa and Glenpool. (Sapulpa Historical Museum).

The Petty and Sloat Livery Company in Kiefer did a prosperous business during the oil boom by renting hacks to oil scouts and lease hounds. Note that while the road is unpaved, wooden sidewalks connect some of Kiefer's businesses. In addition, a wooden driveway has been constructed to allow customers to enter and leave the livery stable without getting in the mud. (Sapulpa Historical Museum).

and in January, 1907, the community was formally platted by the Kiefer Townsite Company. Within two years, Kiefer boasted two banks. By August, 1907, the St. Louis and San Francisco Railroad depot in Kiefer was doing $150,000 worth of business a month, and the "Oil Flyer" special was making daily runs to Tulsa. Two years later, in 1909, D. W. Franchot, constructed the first casinghead gasoline plant in the southwestern United States in Kiefer. Utilizing the output of Glenn Pool, the facility produced "a few hundred gallons" of gasoline every week and brought even more residents to the community.[40]

Because of the frantic pace of drilling in the field, there was no time to store incoming supplies at Kiefer before shipping them to the wells. Equipment thus was unloaded from the trains directly onto wagons. Usually, by 9:00 a.m. every morning there was a two-mile-long-line of vehicles making their way from the depot to the nearby field. The vehicles appeared to one reporter "as...two continuous lines of wagons and teams...[stretched] out as far as the eye can reach, the one string going, the other coming....All the teams coming in are empty while those going out are loaded with all kinds of material for use in drilling, piping and storing oil."[41]

Under such a heavy volume of traffic, the road between Kiefer and Glenn Pool quickly broke down. Deep ruts were cut into its surface and the dirt was ground into a fine powder-like dust that choaked both men and animals. After a rain, the dust was transformed into a thick mud which made travel almost impossible.[42]

One of several fires that destroyed parts of Kiefer during the oil boom. The ramshackle wooden buildings burned easily, and there was little firefighting equipment. This fire, on June 21, 1904, caused $75,000 worth of damage. Note the woman on the right venturing out with an umbrella. (Sapulpa Historical Museum).

However, it was heat rather than rain that caused much of the problems in the summer of 1907. The season was extremely hot and dry, and in July of that year, the temperature remained above 100° for more than two weeks. Under the heat wave, animals quickly tired, and many workers collapsed. Especially hard hit were the tankies working on the large steel storage tanks needed to hold the field's huge output. Working around the hot metal sapped their strength, and, as one witness recalled, "it was a common occurence to see men prostrated while at work."[43]

Combined with the lack of proper sanitation associated with a boom town, the heat created a very unhealthy situation. Hastily constructed shacks housed many workers and their families, and their nearby outdoor privies reeked. With so little rain, the dust settled everywhere and clung to the sweat covered workers until they were caked with dirt. The town soon gained the reputation of being a "hell hole."[44]

Even so, the lure of work and high wages attracted a horde of young men to Kiefer. Most were single. Few of those who were married were willing to bring their families with them to what one described as "A lawless hole of oil field workers." With few of the conventional social restraints in place, many workers spent "their pay with utter abandon" to the benefit of a host of gamblers, prostitutes, bootleggers, and con men in a section of town called the Bowery.[45]

Here a worker could find saloons, brothels, dance halls, and gambling dens. In a place where "feminine society could be purchased" easily, it was not uncommon to see drunk women staggering through the streets "naked and cursing." Because

The unpaved Main Street of Mounds did not dampen the spirit of these marchers; however, the crowds viewing the parade found what shelter they could under the covered fronts of the buildings. Note the floats being pulled by horses leading the parade. (Sapulpa Historical Museum).

Elizabeth Winter's Boarding House in Taneha. The boarding house is actually two shotgun houses joined together. Both structures are nothing but wooden frames with clapboard sidings and tar paper roofs. (Sapulpa Historical Museum).

the oil field operated in shifts, "from one-half to two-thirds of the men" always were at leisure, and thus the "dance halls operated day and night."[46]

Because of the Bowery's reputation, Kiefer was "known as the toughest town east of Cripple Creek, Colorado." The sound of gunfights was heard nightly, and "shooting, knifings, and killings were commonplace" in the district. Jack Dillon, who worked on the nearby oil field, recalled that when he was working with a crew tearing down old oil storage tanks in 1913, the men found the remains of six bodies in one tank, which had been built in 1906.[47]

One of the Bowery's most infamous residents was a saloonkeeper called "Tex," who also "ran a gambling place and sporting houses." Tex was credited by local citizens with "many a killing," and, as Dillon commented, "probably was due the credit." Eventually, Tex's outrages became so flagrant that he was arrested and brought to trial. However, he managed to go free after bribing another man to confess to the charges.[48]

Shobe once visited one of Kiefer's "gambling joints," located across from the railroad depot. "To reach there," he explained, "you had to walk across a narrow plank walk that had been built on stakes across the slough." Gambling and liquor was available downstairs, and upstairs were "the ever-present rooms for the 'girls.'"[49]

The narrow walkway to the gambling house was specifically designed for one purpose. It was only 24 inches wide and anytime someone "made a big killing at the tables" and started home he "was cornered in the middle...by two men coming at him, one from each end of the walk." Trapped in the middle, the winner could do nothing but surrender his money. "It was better than being knocked in the head and dumped into the slough which fairly swarmed with cottonmouth snakes" Shobe explained.[50]

Eventually the violence became too much to ignore and three lawmen were sent from Tulsa to investigate. One of them went to the gambling house and won a large sum of money. However, before starting for home, he positioned the other two lawmen out of sight at each end of the walkway. "Sure enough," Shobe recalled, "as he reached about the middle of the plank walk he saw a man step out on the end in front and he knew there was probably another behind him."[51]

However, this time the other two lawmen rushed up with their guns and captured the hijackers. "With guns in their backs," Shobe continued, "they were marched over to the lights at the depot where their masks were removed." The hijackers were the Kiefer marshal and his deputy.[52]

For three days during the period of transition from territorial status to statehood, Kiefer essentially was without law enforcement. One veteran of the boom later recalled that "so much hell was raised" that no other town could compare to Kiefer in its heyday. One of Kiefer's lawmen, Charles B. Lake, boasted that he was the "only one of five boom town peace officers to survive the experience."[53]

A good example of the disregard for law in Kiefer occurred just after statehood when the neighboring communities of Sapulpa and Bristow were involved in a closely contested election to be named the county seat of Creek County. Because the outcome of the election in either of the two communities was already established, inasmuch as few Sapulpans would vote for Bristow and no one in Bristow would vote for Sapulpa, it was obvious that the outcome would be decided by the citizens of Kiefer. Tempers flared as the election neared. Citizens of both communities began wearing guns.[54]

Hoping to sway the election, a group of Sapulpa citizens offered free whiskey and lunch for Kiefer voters on election day, and as Dillon recalled "we all voted just as long as we could stand up." According to Dillion, "a full vote was gotten out that day" in Kiefer, with many more votes cast than voters registered. With Kiefer's support, Sapulpa was named the county seat.[55]

Eventually, however, Kiefer began to outgrow its boom town atmosphere, especially after many workers began to bring their families with them. With the "respectable women" came a demand for law and order, a demand supported by the local newspaper, the *Kiefer News,* which exposed the community's more obvious vices. Also with the families came the need for churches and schools. Although initially many churches were forced to hold services in private homes or under brush arbors, ultimately church buildings were built. Likewise, responding to pressure to offer adequate educational facilities, local officials constructed a high school that boasted of being the "most modern and well equipped" in Oklahoma. When the Bowery's worst offenders opposed the "civilizing"

Sherman Smith's gasoline service station in Taneha. Note the concession stand to the right of the gasoline station offering Budweiser Beer and Grape Bouquet soda. (Sapulpa Historical Museum).

of the community, Kiefer's more respectable citizens formed a vigilance committee and drove them out of town. With their departure, the transformation of Kiefer into a modern progressive community was completed.[56]

As oil production dropped in the immediate vicinity, so did Kiefer's population. The declining number of citizens no longer could support the size of business district that had developed during the heyday of the oil boom, and many businesses failed or moved to another boom town. The decline was so drastic that by 1940 Kiefer had only 330 residents. By the early 1950s its population had decreased to approximately 200. The trend finally reversed as the town became a commuter suburb of nearby Tulsa. By 1980 Kiefer's population was 1,086.[57]

Two other oil boom communities near Glenn Pool were Taneha, or Bowden, about four miles north of Sapulpa and Sabo, two miles north of Kiefer. Both were "rough oil field" communities consisting of clapboard buildings and tents. Taneha boasted a grocery store, hotel, and a few other wooden structures along a single street. It was not, as one resident explained, a town in which women "get outside after dark."[58]

Sabo did not last long after the Glenn Pool oil boom. Its post office opened in 1907 and closed in 1913. Bowden, however, existed for almost half a century. Its post office opened at the height of the oil boom in 1909, and did not close until 1957.[59]

About eight miles south of Kiefer, along the St. Louis and San Francisco right-of-way in southeastern Creek County, was a small community originally known as Posey. However, just to the west of the townsite Duck Creek flowed past a very prominent hill or mound, which was easily identifiable. As a result, the town's name was changed to Mounds on April 19, 1898.[60]

Throughout its early history, Mounds, like other similar communities in the region, was a small agricultural settlement serving area farmers and ranchers. However, after the opening of Glenn Pool, The Texas Company selected Mounds as its

Delivery truck of the Sabo Grocery. The vehicle, driven by Harry Young, made frequent deliveries of groceries to the homes and oil field camps surrounding Sabo, located approximately two miles north of Kiefer. (Sapulpa Historical Museum).

major supply point for the nearby field. In addition, the St. Louis and San Francisco Railroad built two large oil loading racks at Mounds for Glenn Pool's crude. Eight trains a day stopped at Mounds. In addition, the community was connected to Tulsa by an interurban, which had hourly runs.[61]

Because it did not have the same boom town reputation as Kiefer, Mounds became the headquarters for several prominent oilmen. Among them were John Rhodes, Buck Self, Joe Berryhill, Hazel Vowell, Matt Rucker, W. H. Milliken, S. W. Anthony, and Barney Flynn. In 1938, Anthony was reported to have paid the largest individual income tax in the state. In addition to his home, Millikin constructed an oil storage facility in the town to handle his Glenn Pool leases.[62]

Ten miles south of Tulsa, between Tulsa and Glenn Pool was Jenks. Named for a local citizen, Elmer E. Jenks, a post office was established in the community in January, 1905. Like Kiefer, Jenks owed its existence to the oil strike. Located along the Midland Valley Railroad right-of-way, it became a major supply and shipping center for the field, especially after the autumn of 1906, when the Midland Valley extended a spur line from Jenks to the town of Glenpool.[63]

Jenks also became a major storage center for much of Glenn Pool's crude and was the site of the Prairie Oil and Gas Company's tank farm. By October, 1907 the town was surrounded by more than 120 huge steel storage tanks. To ship its crude, the Prairie Oil and Gas Company constructed a half-mile-long loading rack on the Midland Valley at Jenks. The facility, which could accommodate 27, 250-barrel tank cars at a time, was so busy that "long lines of tank cars could be seen loading and waiting to be shipped." It was from Jenks on March 30, 1907, that the Prairie Oil and Gas Company shipped the first train load of Glenn Pool oil to market, and by September, 1907, the firm was dispatching 245,000 barrels of Glenn Pool crude out of Jenks every month.[64]

Because of the many nearby storage facilities, Jenks was the home for many of the tankies who built the tanks and the workers who maintained the shipping facilities. Because jobs at the tank farm were more enduring then jobs with a drilling crew, many of the workers in Jenks brought their families with them. As a result the community never acquired the same reputation as a boom town as Kiefer.[65]

Associated with Jenks were the nearby oil field camps of Watkins and Poag, which were located on the spur line between Jenks and Glenpool. The site of the Gulf Oil Company's loading racks, Watkins was the center of Gulf's gathering lines running throughout Glenn Pool. The crude accumulated at Watkins, was loaded into railroad tank cars and sent to Jenks for transshipment to other markets.[66]

While Glenn Pool produced one of Oklahoma's most infamous oil boom towns—Kiefer—it stimulated the growth of several other nearby towns that became respectable communities. Jenks, Mounds, Glenpool, and Sapulpa, today are thriving modern communities. Likewise, Kiefer has successfully made the transformation from a boom town to a quite surburban community. While all five towns still have strong oil ties and serve as the base for several local oilmen, they all eventually were overshadowed by Tulsa—"Oil Capital of the World."[67]

CHAPTER NINE

TULSA: OIL BOOM TOWN

Between the 1830s and the 1890s the Creek name for the settlement on the north bank of the Arkansas River, Tallahassee, had been corrupted first to Tallasi and then Tulsi Town by non-Indians. Although it already had been a settlement for more than half a century, it was not until the allotment of Creek lands by the Dawes Commission that the community, officially named Tulsa, was formally incorporated on January 18, 1898. Within two years of its original platting, Tulsa stretched for a block along both the east and west sides of Main Street, along Boulder and Boston avenues, and south of the St. Louis and San Francisco Railroad tracks for five blocks along First, Second, Third, Fourth, and Fifth streets.[1]

The main business district was located between Second Street and the railroad. On the north side of the tracks were the Brady Store, the St. Elmo Hotel, and the Dickerson Lumber Co. To the south were Gill's Blacksmith Shop, the J. M. Hall Store, the Lon Stansbery Implements establishment, the Spence Kellum Livery Stable, Jack Archer's Store, the Bynum Store, the Dave Shultz-Dry Goods establishment, Williamson Brothers, W. O. Struther-Drugs, Dr. J. M. Morrow-Drugs, Hogan's Livery, Lonnie Price Harness, the Lynch Brothers' General Store, the Tulsa Banking Company, Linsky Tailor Shop, Winkler Harness, the Club Hotel, Edgar Brothers' Building, Price and Gillette-General Store, Wallace and Calhoun-Meat Market, Pat Coyne Drugs, and Fred Scott Dry Goods. On the east side of Boston Avenue, between Second and Third streets, was a pool hall, and two blocks south of there, between Fourth and Fifth streets, was the school. The community boasted three churches—the First Christian, the First Methodist, and the Presbyterian Mission—and a newspaper—the *Indian Republican,* later the *Tulsa World.* By 1901, Tulsa's population had reached 1,370, ranking the community 31st according to size in Indian Territory. The rural lifestyle of Tulsa changed drastically, however, after the discoveries at Red Fork and Glenn Pool, as thousands of oilmen rushed to the town.[2]

Many were lured by stories of vast fortunes. R. A. Josey, for example, who arrived in Tulsa in 1906, immediately began investing in the Glenn Pool region. A year later, in February, 1907, Josey was a well-known independent oilman worth hundreds of thousands of dollars. The profits on his early investments were astounding. One 50-acre lease, purchased for $600 in September, 1906, brought Josey $70,000 four months later. Another 160-acre lease, which had cost $500 in May, 1906, was sold by Josey in February, 1907, for $80,000.[3]

To handle the huge influx of oilmen rushing to Tulsa after the Ida Glenn No. 1, the St. Louis and San Francisco Railroad ran three trains daily from St. Louis and another four trains daily from Kansas City to Tulsa. Advertised as "the best trains to the Tulsa Oil Fields," each was equipped with Pullman sleeper cars for those who could afford to travel in luxury. Meals, prepared by Harvey House personnel, were available in the dining car.[4]

Many of the early arrivals in Tulsa were able to secure lodging in the St. Elmo, Club, and Alcorn hotels or Jane Perryman's Boarding House. Shortly after the discovery at Glenn Pool, another hotel, the Cottage Home Hotel or Col. Moore Hotel, was opened to handle the influx of people. Boasting of "hospitality...[for] all who have stopped in the city to refresh the inner man," the Cottage Home offered "Rest and refreshment" for at a cost of $3.50 weekly or $2.00 per day.[5]

Within a year of the strike, so many oilmen had arrived in Tulsa that housing virtually was impossible to find. Many were taken into private homes, but these too were quickly filled and men were forced to sleep on derrick floors. Hurriedly constructed buildings were built to provide shelter, but many were nothing more than "flee bags, always overcrowded, [and] full almost every night." Generally, they offered "small rooms, hard beds and neither heat or stands for pitchers of water."[6]

The dearth of proper housing especially was noticeable in the scarcity of adequate bathing facilities. Ordinary washtubs were sold as bathtubs so that oil field workers could bathe. Even so, a long wait in a line usually was necessary before

Looking south from the railroad depot along Tulsa's Main Street in 1896. Prior to the oil boom, Tulsa was only one of several rural agricultural communities serving the region. (Beryl D. Ford).

one could find a bath. To solve the problem, some oilmen brought their own bathtubs with them and hired young boys to carry water to their rooms.[7]

However, not all of the newly constructed hotels were typical boom town shanties. William H. Robinson built the luxurious Robinson Hotel. Located at the intersection of Third and Main streets, it was 5 stories tall and contained 146-rooms. Catering to the more affluent, the Robinson Hotel quickly became the center of oil field action in Tulsa. Boasting "clean sheets, good food, and speedy service," it was recognized as one of the finest facilities between the Mississippi River and the Rocky Mountains. Harry F. Sinclair, who at one time controlled the largest independent oil company in the world, often conducted business while walking between his room and the lavatory in his pajamas. As one long-time oilman pointed out, "more wells were drilled in the Robinson Hotel lobby than in the Glenn Pool field."[8]

The Hotel Tulsa, at Third Street and Cincinnati Avenue, was another of the community's more famous accommodations. After it opened on May 12, 1912, it quickly acquired a reputation as the "best place to stay, to lounge, to eat, to get information, and to give away information" on the oil patch. Lease hounds were everywhere hunting for "dope" as they called information on the latest strike. Some oilmen made the Hotel Tulsa's lobby their office, conducting business from the huge leather chairs provided for patrons. So that he could be near the scene of the latest action, Sinclair maintained a suite of offices on the fifth floor of the Hotel Tulsa, and Joshua Cosden, who once wrote a $12 million check in the hotel's waiting room, was a frequent visitor. It was estimated that more than a billion dollars changed hands in the hotel's lobby.[9]

While many leasehounds traded property in the hotels, it often was necessary for them to actually visit the surrounding oil fields to get accurate information. To get there they had to make their way through the "piles of construction material [that] littered...[the] streets" toward the

The same view south from the railroad depot along Tulsa's Main Steet in 1907. By this time, the nearby oil strikes at Red Fork and Glenn Pool had stimulated the growth of the community, but the street remains unpaved. The interurban line clearly is visable. Note the outlet of the Oil Well Supply Company to the right of the photograph. Next to it is a hotel offering food and lodging to oilmen. (Beryl D. Ford).

An interurban car on the line between Tulsa and Glenn Pool. The interurban allowed many oilmen to live in Tulsa, where they were not exposed to the more rowdy elements of the oil boom, and commute by rail to their jobs. (Sapulpa Historical Museum).

Looking south from the railroad depot along Tulsa's Main Street in 1909. By this time the street has been paved with bricks. (Beryl D. Ford).

railroad station, where many paused at the Pig's Ear, a resturant located just across the street from the depot, to eat breakfast and purchase a box lunch for a mid-day meal. Then they would catch the "Coal Oil Johnny," a special 15-car train that carried them to the oil fields. That evening they caught the same train for the ride back to Tulsa.[10]

Like most other oil boom towns, Tulsa had its more rowdy side. "Gambling houses opened as rapidly as second-story locations were made available," one oilman recalled, while crap tables and roulette wheels were set up in "clubs." One of the most notorious establishments was the Bucket of Blood Saloon, where "fights were had about every night" and "several murders occurred." Other, less visible vices were housed in tents and shacks throughout the community's business district. Hiawatha, Zenol, Jik, and Ino and Uno beer were popular, as was homemade beer. Bootleg liquor was plentiful.[11]

Prior to statehood, Tulsa was officially dry; however, it was only a few miles to Oklahoma Territory where liquor was legal. Keystone, the first settlement west of Tulsa in Oklahoma Territory, became a major supply point for illegal whiskey. The "Keystone Trail," a major route used by liquor smugglers, connected the two communities. If the overland route became too dangerous because of federal agents, the whiskey runners often turned to the Arkansas River, when its water was sufficiently deep. They simply loaded rafts with whiskey legally purchased in Keystone and followed it along the shore as it floated downstream to Tulsa. Should they be approached by lawmen, they claimed to be innocent fishermen.[12]

Few people made any attempt to halt the illegal flow of liquor. Yet, on one occasion word reached Carry Nation, the ardent prohibitionist, that Tulsa's Robinson Hotel planned to celebrate its opening with "a large quantity of real beer and some hard liquor." "Carrying a hatchet," and, as one witness described, "looking like a battle-ax," she hurried south from her home in Atchison, Kansas, determined to put an end to such an open violation of the liquor laws. Speaking before Tulsa's Commercial Club, Nation denounced the hotel's "evil" plans and proclaimed that if local lawmen would not put a stop to the illegal activity, she was

A view of the tank farm of the Cosden Refinery in West Tulsa. The active solicitation of oil companies by Tulsa's businessmen greatly contributed to the tremendous amount of capital which poured into Tulsa during the Red Fork-Glenn Pool oil boom. Note that several sites in the center of the photograph have been prepared for the construction of 55,000-barrel oil storage tanks. (PennWell Publishing Company).

there "to stamp…out" the bootleggers. If "her word wasn't heeded," she warned "she'd cut up the Robinson's dining room."[13]

After appearing before the Commercial Club, Nation walked along Tulsa's streets passing out "souvenir hatchets" to those that would take them. Within a short time of her speech, word spread throughout Tulsa of her threat. When officials of the hotel heard of her warnings, they prudently moved the time of the celebration forward 30 minutes and announced "that everyone was expected to be prompt" if they wanted to be admitted."[14]

As a result, when Nation arrived to put an end to the "debauchery," she found the doors to the dinning room "securely locked and bolted" While she "pounded loudly, gloriously, [and] vainly" on the door, 200 guests enjoyed their meal and toasted her crusade. Standing outside, Nation could clearly hear the toast and the beer bottles clinking. Indignant, she wasted little time in boarding a train back to Kansas to leave "the heathens to rot in hell."[15]

Tulsa's oil boom never got completely out of hand as it did in Kiefer. Almost as soon as the discovery was made at Glenn Pool, the community's civic leaders took measures to control the inevitable vice and violence that followed oil booms. John A. Oliphant, who arrived in Tulsa in 1905 and became the community's first Police and Fire Commissioner, led the fight against the more rowdy elements, especially those "living in sin." "I found there were a good many men—oil prospectors for the most part—living rather careless domestic lives," Oliphant explained. "Some of them had left wives in their far-off eastern or northern homes" and as a result "had formed misalliances here." In fact, he pointed out, "the number of housekeepers, cooks and laundresses they had was amazing."[16]

Such illicit relationships were not proper in Oliphant's eyes. "I thought it wouldn't do" he explained. "The conduct of these men and their women friends was growing flagrantly bold….So I determined to clean up the Augean [i.e., filthy] stables, so to speak," the Commissioner continued.[17]

"I went about it in the most practical way I knew," Oliphant explained. "I simply inserted an official advertisement in the two newspapers warning these men either to marry their housekeepers, laundresses, cooks, etc., or do their own cooking, laundrying and housekeeping." As a result, the Commissioner boasted, "many of them left the city."[18]

In another effort to eliminate much of the vice associated with the oil boom, a meeting was held, in February, 1910, at the Bijou Theater to organize a "law and order" slate of municipal officials to oppose the incumbent, Mayor John O. Mitchell, in the coming elections. After listening to L. J. Martin and J. M. Hall speak on the advantages of ridding the town of undesirables, the gathering selected Hall as chairman of a nonpartisan

The Prairie Oil and Gas Company's Tulsa warehouse crew. Tulsa quickly became a major supply center for the surrounding oil fields. The Prairie Oil and Gas Company facility contained a complete machine shop operation, which enabled the warehouse to make many parts on demand. "Old Bird," the crew's mascot, is third from the right on the front row. (PennWell Publishing Company).

group pledged to an anti-vice campaign. He was joined on the committee by C. L. Holland, C. J. Sawyer, I. J. Buck, F. M. Roddy, and R. E. Lynch, while H. O. McClure was named the group's candidate for mayor.[19]

However, L. J. Martin deserted the nonpartisan coalition and announced his own candidacy as a democrat, and he won the election. Almost immediately, he hired Herman Newblock, who Martin claimed had been defeated for re-election as Tulsa County Sheriff by "the gamblers in town" to head the city's police. Martin and Newblock's announced goal was to "drive the open saloons and the open gambling out of the city."[20]

Newblock tackled the task in earnest. He raided the saloons, arrested gamblers, and closed down bootleggers. To circumvent Newblock's tactics, the saloonkeepers and gamblers simply moved just outside the city limits and reopened. Moreover, they hired female bartenders and offered free rides from Main Street to their establishments.[21]

Martin was succeeded as mayor in 1912 by John N. Wooden. According to his opponents, Wooden was "weak on law enforcement," and during his administration they claimed that much of the vice returned to the city. Unpopular, Wooden became the object of several sermons by local preachers, with one, the Reverend Percy A. Knickerbocker of the Boston Avenue Methodist Church, accusing the city's officials of finding it "unpopular to enforce the law" and of turning Tulsa into a mecca for "under-world characters...from Oklahoma City and Muskogee."[22]

In spite of such opposition, Wooden was reelected in 1914. During his second administration, however, the corruption became too obvious for many citizens. In response, a Law Enforcement Club was formed, with H. C. Tyrrel, B. W. Grant, and Arthur Hull named as Chairmman of Publications, Secretary, and Treasurer respectively. Other members of the organization's Board of Directors were Murray M. Doan, E. M. Arnold, Carl C. Magee, H. W. Randolph, J. B. Wilson, C. A. Sanderson, P. L. Price, and A. F. Ault.[23]

Instead of direct vigilante action, the Law Enforcement Club published *The Daily Independent,* a weekly newspaper with the motto of "a very little paper with a great big mission." On its pages were the names of many Tulsa area underworld leaders and municipal officials who were suspected of taking bribes. Claiming the community contained "at least ten big open gambling houses" in which "from 50 to 300 men nightly" congregate to "gam-

Headquarters of The Texas Company's West Tulsa Refinery complex. Note the big red star with a T on the top of the building, the Texaco symbol of today. (Sapulpa Historical Museum).

ble and drink and carouse," the newspaper printed the names and address of the worst "joints."[24]

More drastic action was taken after two United States marshals were shot and killed on July 2, 1914, while raiding the home of a known bootlegger, who was a former chief of police. Irate over the killings, a mass meeting of Tulsa citizens, many of them members of the Law Enforcement Club, gathered on July 3 at the corner of Fourth and Main streets. After listening to accusations of a "lawless Tulsa," the crowd called upon Governor Lee Cruce to help rid the city of its unsavory elements.[25]

A grand jury subsequently was called, and Wooden eventually was indicted and removed from office. His post was filled by O. D. Hunt. Frightened by the grand jury indictments, corrupt city officials refused to continue condoning much of the previous vice.[26]

Many of the pioneering oilmen who rushed to Tulsa after the nearby strikes became well-known characters within the profession. Colorful, and often exuberant, some spent their new-found wealth with abandon. Two in particular were W. F. (Billy) Roeser and Joshua S. Cosden. Both made millions of dollars in the Oklahoma petroleum industry, became instant celebrities, and died impoverished.[27]

Born into an oil family in Marietta, Ohio, Roeser moved first to Beaumont, Texas, where he was employed by the Texas District Petroleum Company, and then, in 1904, to Muskogee. After two months in Muskogee, Roeser migrated to Cleveland and then to Tulsa. Although he was broke when he arrived, he rented an office in Tulsa's First National Bank Building and managed to raise sufficient capital to form several oil companies—the Oklahoma State Oil Company, the Ethel Oil Company, and the Coast Oil Company—which operated in Glenn Pool.[28]

Within a short time, Roeser's Glenn Pool holdings were producing a $3,000-per-day profit. He spent the money as fast as he received it. As one acquaintance pointed out, Roeser "loved to drink,

Tulsa's skyline in 1939. By the close of the first three decades of the 20th century, Tulsa, the oil boom town, was hailed as Tulsa, the oil capital of the world. The Mayo Hotel on the right was the headquarters for many early-day oilmen and millions of dollars exchanged hands in deals struck in the hotel's lobby. (Oklahoma Historical Society).

drive recklessly and thumb his nose at anyone who sought to pass judgement." True to his fast style, he bought the former home of George Byrne at 1437 South Boulder. One of Tulsa's largest mansions, it cost Roeser $35,000. However, the price of the building was only a portion of Roeser total expenditure.[29]

Because all the furniture that came with the house was not mahogany, Roeser disposed of it and hired a firm of interior decorators from St. Louis, Missouri, at a cost of $65,000, to redo his new home. Another $16,000 was spent for works of art to grace the interior. Fifteen railroad freight cars of blue grass from Neosho, Missouri, was ordered to landscape the two-acre yard at a cost of $7,000. An additional $8,500 was spent for trees and shrubs. When the redecorating was completed, Roeser spent $5,000 on a housewarming party.[30]

Roeser also liked fast automobiles, and his garage usually held 6 cars, worth about $16,000. It was a common sight to see him roaring through town. Indeed, he thought so little of spending money, that when one of the automobiles became stuck in the mud, he sold it for $200.[31]

Roeser's luck finally ran out. After giving away vast amounts of his fortune to friends and spending huge sums of money to finance wildcat wells, his capital began to run out. The final blow came when he acquired 110,000 acres of prospective oil properties in Illinois that produced nothing but dusters. Suddenly Roeser found himself broke.[32]

However, he refused to quit. When he told his family of his losses, he remarked "Look. Here are two 20 dollar bills. That's what I had before I started. So there's no way to go but up." Although he lost his home, his works of art, and his furniture, his friends advanced him sufficient credit to start again.[33]

With their help, Roeser made a mild comeback and recovered part of his lost fortune. However, he continued to gamble and invested it all in a 20,000-acre-lease near Maramec, Oklahoma. When asked what he would do if it proved to be non-productive, Roeser replied, "so what? That's the game. Win a few. Lose a Few...in this business of oil, a man's got to be game. Risk it all for a big stake." It proved to be worthless, and he was broke again. All total he had gone through $50 million.[34]

Like Roeser, Joshua S. Cosden, who pioneered the effort to make Tulsa a refining center, made and lost two fortunes. "Handsome, intelligent, charming, and generous," Cosden, was known as the "Prince of Petroleum" when he made his first fortune in the Osage Nation near Bigheart. Using his Osage money to finance the project, Cosden built a refinery complex in West Tulsa which became one of the largest independent refineries in the world. Eventually his personal wealth grew to $50 million. He later built the Cosden Building, which was renamed the Mid-Continent Building, Tulsa's first skyscraper.[35]

At the height of his wealth, Cosden went on a spending spree. He divorced his first wife, and then married Mrs. Charles Roeser, Billy Roeser's ex-sister-in-law. Cosden and his new bride purchased an estate on Long Island, New York, and another in Palm Beach, Florida. He also acquired a suite in the Plaza Hotel, a private racing stable, his own railroad car, a yacht, and a 100,000-acre ranch in Florida. His extravagance knew few bounds, and he once presented his wife with a $600,000 necklace. By 1925 he was broke.[36]

Within two years however, Cosden, with the

backing of New York financiers, had regained $15 million of his lost furtune in the Texas oil fields. He lost most of it in the Great Depression and died impoverished in 1940. Nonetheless, Cosden's former businesses dominated the industrial life of Tulsa for many years.[37]

In spite of its oil boom, however, Tulsa never acquired the unsavory reputation of many of Oklahoma's oil communities. It did not have the violence of Cromwell or the vice of Wirt. Nor did it disappear when the strike was over. Instead, Tulsa outlived the boom to become a major oil center and claim the title of "Oil Capital of the World."[38]

CHAPTER TEN

TULSA: OIL CAPITAL OF THE WORLD

During the heyday of the Red Fork, Cleveland, and Glenn Pool booms, Tulsa grew beyond the typical oil field settlement to become the "most significant and outstanding oil boom town of the Southwest." The reason for this metamorphosis was the early realization by Tulsa leaders of the economic potentials of the nearby oil strikes and their coordinated efforts to entice independent oilmen and major companies to the community. Because of such an attitude by business and civic leaders, Tulsa soon was being hailed as the "Oil Capital of the World."[1]

Tulsa's leaders wasted little time in promoting the community's prominence in the petroleum industry. One of the earliest examples of such boosterism was a flaming gas well at the intersection of Seventh Street and Boulder Avenue, which greeted visitors to the town. The gasser had been drilled shortly after the completion of the Sue A. Bland No. 1. Inasmuch as the natural gas practically was worthless, local boosters attached to the well head a pipe that extended several feet into the air. Then a "tin-can" was placed over the exposed end of the pipe and the gas was set on fire to make a flare.[2]

With the flames jumping 30 feet into the air, the flare became quite a center of attraction and caught the eye of most newly arrived oilmen. If the boosters wanted to impress a certain group of potential investors, they would turn off the gas flow to allow the pressure to build up. Then after the new arrivals had gathered around the well head, it would be turned on with a roar. Of course some of the potential investors invariably would stand too close to the pipe and be startled by the flames or even singed. One remarked, as the flames burst out of the pipe, "Hell is only three feet from here."[3]

In 1902, Tulsa businessmen and civic leaders formed a Commercial Club. A forerunner of the Chamber of Commerce, it was dedicated to the philosophy of making Tulsa a major oil center, and as by 1905 was touting the community as a place where "men of vision were always welcome in the councils of the great." To present a more positive image of Tulsa to neighboring states, the group organized "booster special" trains that visited surrounding areas promoting the town and urging oilmen to make it their headquarters.[4]

The trips became quite sophisticated and by 1908 consisted of five railroad cars: a day coach, three Pullman sleepers, and a baggage car, which held a printing press to publish a daily newspaper to distribute along the route. Annie Oakley headlined the trip that year, along with about 125 Tulsa boosters. Among their stops were Chicago, Illinois; Cincinnati, Ohio; Washington, D. C.; and Baltimore, Maryland. In 1919 another International Special made a 4,000-mile-long trip through much of the United States and into Canada.[5]

The 1905 trip, however, proved to be the most famous of all. When the Tulsa boosters paused in Terre Haute, Indiana, a local brewery had filled the train's baggage car with beer and afterward the trip took on a new level of gaiety. Most of the boosters were full of the "spirit" by the time the train reached Chicago, and thus when officials of the Pullman Company attempted to remove the banners draped from the train's cars, they incurred the wrath of the more vocal boosters. Unable to remove the banners, which Pullman rules prohibited, company officials called the police. In the meantime, a large crowd had gathered. While the boosters and the police were locked in a standoff, the boosters' band climbed on top of the cars to play for the crowd. Eventually a compromise was reached, and the boosters agreed to remove the banners while the train was in Chicago. Afterward, as one Pullman official proclaimed, "you can put them back on and run the train to hell and back if you want to."[6]

The boosters were successful, and from the beginning of the Oklahoma petroleum industry many oil leaders established their headquarters in Tulsa. A major reason for their choice of Tulsa was its less rowdy atmosphere and more readily available housing. They in turn attracted a large number of supply houses and machine shops whose salesmen hoped to do business with them. Likewise, with not as much vice and violence as

The First National Bank of Tulsa, ca. 1902, was only one of several oil-related banks in Tulsa that provided much of the capital needed to develop the surrounding oil fields. Such an active policy of support by local banks freed Tulsa's oilmen from dependence on Eastern capital to finance their operations and did much to make Tulsa the oil capital of the world. (American Petroleum Institute).

other boom towns, people with families were willing to settle in the community. This in turn provided a ready pool of stable workers for the refining, marketing, and office aspects of the oil business.[7]

In addition to Tulsa's attractions in accommodations and work force, its location and adequate rail and pipeline service played a prominent role in the town's development as a major oil center. Within a radius of 100 miles from Tulsa were several major oil strikes—Glenn Pool, Cushing, the Osage—and literally hundreds of smaller discoveries. Running north and northwest toward the Osage were the Midland Valley and the Atchison, Topeka and Santa Fe railroads. Stretching south and southwest toward Cushing, Glenn Pool, and Red Fork were the St. Louis-San Francisco and the Missouri-Kansas-Texas railroads. In addition, the Sinclair, Cosden, Pierce, Oklahoma, Gulf, Texas, Empire, Magnolia, and Prairie pipeline systems converged on Tulsa. As a result, Tulsa's growth was stimulated by numerous nearby oil fields to which it was connected by rail or pipeline networks.[8]

Another important early development contributing to Tulsa's rise as the "Oil Capital of the World" was a strong pro-oil banking establishment. In the early-days of the Oklahoma oil boom, the tremendous amount of capital necessary to finance the development of a major oil strike usually came from the East. This, of course, made expansion of Oklahoma's oil industry dependent upon the judgments of conservative Eastern financiers. This practice was altered in 1910 when Harry F. Sinclair, Pat J. White, J. H. Evans, F. B. Ufer, James A. Chapman, and Robert M. Mc-

It was a short ride from the Kiefer depot, shown here, to downtown Tulsa. The availability of reliable transportation to the nearby wells encouraged many oilmen to make Tulsa their headquarters. (Sapulpa Historical Museum).

Farlin acquired control of Tulsa's Farmers National Bank and reopened it as the Exchange National Bank with a capitalization of $400,000.[9]

Strictly an oil bank, it replaced the region's total dependence on Eastern capital with a local source of financing. No longer did a local oilmen "have to wait on the impersonal action of Eastern financial centers." Instead, "he got immediate and decisive action," for he was served by an official entirely familiar with the prospects and possibilities of his lease and the value of any security he might offer as collateral.[10]

Within seven years of its acquisition by Sinclair and the others, the Exchange National Bank claimed assets totaling $21 million and by 1920 was doing business world-wide. Its growth was phenomenal and by 1929 the bank occupied a 29-story building and boasted assets of $50 million. The Exchange National Bank failed during the Great Depression, despite the efforts of several Tulsa oilmen including James Chapman who offered $5 million to keep the bank open. Even so, it was reopened in 1933 as the National Bank of Tulsa, eventually became the Bank of Oklahoma, and continued to play a major role in the petroleum-natural gas development of the state.[11]

Supporting the business community, Tulsa's civic leaders, early in the oil boom, decided to make the community the region's economic center. The initial example of this attitude came even prior to the Glenn Pool discovery, when Tulsa's only link to the southwest was the Frisco Railroad. Other overland routes stopped at the banks of the Arkansas River. Because of the physical problems of getting from Tulsa to the well sites, many Tulsa businessmen feared that other, more convenient towns would become the sites for area refineries and company headquarters. Obviously a bridge over the Arkansas River was a vital necessity, so that oilmen could make their way to Red Fork or Glenn Pool at any time rather then being forced to depend on the railroad schedule. Realizing this, Tulsa's leaders proposed a bond issue to finance the needed bridge, but the voters failed to approve the issuance of bonds. The only alternative was a private venture.[12]

Patrick C. Boyle, president of the Petroleum Publishing Company from 1910 to 1920, moved the publication of the Oil and Gas Journal from Beaumont, Texas, to Tulsa, thereby lending credence to the community's claim as the oil capital of the world. (PennWell Publishing Company).

Waite Phillips, one of many pioneering oilmen who made their home in Tulsa. (PennWell Publishing Company).

Despite warnings by engineers that the sand in the river bottom was not sufficiently strong to support such a structure, George T. Williamson, J. D. Hagler, and M. L. Baird secured a franchise from the federal government and began work on what became the Eleventh Street Bridge. Old boilers were filled with concrete and lowered into the water as piers, and nearby trees were cut to provide limber for the floor. Costing $50,000, quite a sum in the early 20th century, it opened in January, 1904.[13]

The push toward achieving economic dominance of the area became even more intense after Glenn Pool. At the time of the completion of the Ida Glenn No. 1, the nearest refinery available to Tulsa area oilmen was at Neodesha, Kansas. Although it paid $1.00 per barrel for crude, the cost of transportation from Red Fork or Glenn Pool was approximately 90¢ per barrel. With a profit margin of only 10¢ per barrel to cover other expenses, the oilmen obviously needed a refining complex in the immediate area. Tulsa's leaders also recognized that the town that controlled the refining industry, and not necessarily the town closest to the strikes, would become dominant.[14]

To insure that Tulsa became the refining center of the region, Tulsa's Commercial Club joined in a project in 1906 to construct the Uncle Sam Refinery in West Tulsa. Although the project failed, their activity attracted the attention of The Texas Company, now Texaco, and in 1910 it opened a 3,000-barrel-per-day refinery in the same area.

Built by Waite Phillips, Tulsa's Philtower Building dominated the community's skyline during the heyday of Oklahoma's oil boom. (Beryl D. Ford).

Harry H. Rogers, another of Tulsa's early-day leaders of the petroleum industry. (PennWell Publishing Company).

Before closing in the 1980s, it had grown into a 466-acre refinery complex that was capable of handling 50,000 barrels every 24 hours.[15]

In 1912, Joshua S. Cosden opened the Cosden Refinery, which occupied nearly 800 acres in West Tulsa. Initially Cosden transportated crude to his refinery in a "battered old tank wagon" which leaked so badly as it "creaked and ground its way" from the wells to the complex that he was forced to follow behind the wagon with a bucket to catch the leaking oil and pour it back into the tank. From this primitive beginning, however, the facility was expanded until it could refine 90,000 barrels of crude daily. Eventually the Cosden Refinery became a part of the Sun Petroleum Products Company, a subsidiary of the Sun Company. Another Tulsa based refinery, the Water-Pierce Oil Company facility, was opened in West Tulsa in 1913.[16]

Tulsa's claim as the "Oil Capital of the World" was solidified in 1908 when *The Oil Investors Journal* relocated in the community. First printed in Beaumont, Texas, on May 24, 1902, by Holland S. Reavis, the publication quickly became recognized as a leading source for oil industry information. In fulfilling its promise to carry "reports from the field, economic analysis, and statistics on drilling, stocks, shipments, and prices of crude oil and refined products," *The Oil Investors Journal* rapidly became known as the "Oil Man's Bible."[17]

When Glenn Pool boomed, *The Oil Investors Journal* opened a branch office in Tulsa. Later, on April 20, 1910, the magazine was sold to the Petroleum Publishing Company owned by Patrick C. Boyle, a Pennsylvania newspaperman who had built a reputation as publisher of the *Oil City Derrick*. The sale price was $20,000.[18]

Boyle quickly changed the name to the *Oil & Gas Journal,* and moved the entire operation to Tulsa; however, he continued to maintain his home in Pennsylvania. Tying the magazine even more closely to the oil and gas industry by providing more current information, Boyle issued the *Journal* weekly, instead of bi-monthly.[19]

When Boyle died in 1920, his son-in-law, Frank T. Lauinger, who continued to reside in Pennsylvania, assumed control of the business. Lauinger "opened the *Journal's* pages more widely to the rapid advances in petroleum technology" and made the magazine a world-wide organ. He was succeeded in 1931 by his son, P. C. Lauinger, who

P. C. Lauinger, second from the left, oversaw the publication of the Oil and Gas Journal from 1931 to 1969. Often called the bible of the oil industry, the publication was relied upon for the most up-to-date oil and gas information. (Interstate Oil Compact Commission).

actually moved the family home to Tulsa. In 1969, P. C. Lauinger became Chairman of the Board of the Petroleum Publishing Company, and the day-to-day running of the business was turned over to Phillip Lauinger, Jr., the fourth generation of the family to oversee the magazine's operations. Eventually the name Petroleum Publishing Company was changed to PennWell Publishing Company.[20]

Because of its rapid and continuous rise in prominence as an oil center, Tulsa became the home for numerous well-known oilmen. People such as J. Paul Getty, founder of Getty Oil Company; W. G. Skelly, the organizer of Skelly Oil Company; James A. Chapman and Robert M. McFarlin, the major operators of the McMan Oil Company, the largest independent firm in the world dealing strictly in oil production in 1916; Waite Phillips, who built the Philtower, Philcade, Beacon, and Elliott buildings; David R. and S. Miller Williams, founders of The Williams Companies, and many others made their home there. Even Dr. Fred S. Clinton, one of Red Fork's pioneer oilmen, owned a home at the corner of Fourth Street and Boston Avenue. And Harry Sinclair eventually moved to Tulsa in 1915 and built a mansion at 1730 South Cheyenne that occupied the entire block.[21]

While all of these pioneering oilmen contributed to Tulsa's rise as the "Oil Capital of the World," Skelly deserves much of the credit for making Tulsa the center of the international petroleum busi-

Skelly Oil Company's Tulsa headquarters. Skelly Oil, founded by W. G. Skelly and now part of the huge Getty Oil network, was but one of many Tulsa-based petroleum concerns. (Getty Refining and Marketing Company).

ness. Not a native Oklahoman, Skelly had grown up among the wells surrounding Oil City, Pennsylvania, and then moved west after serving in the Spanish-American War. He migrated first to the oil fields of Ohio, Kansas, and Texas before moving to Oklahoma's Healdton-Hewitt Field, where he worked for Wirt Franklin. In 1917, he constructed the Midland Oil Refinery in El Dorado, Kansas, and this, combined with his profits from investments in the Healdton-Hewitt Pool, provided sufficient capital for him to join Jack Sankey in forming Skelly-Sankey Oil Company that same year.[22]

Two years later, on October 2, 1919, he formed the Skelly Oil Company. Based at Fourth and Boulder in downtown Tulsa, but incorporated in Delaware, Skelly Oil Company was capitalized at $15 million. Half of the money was represented by the assets of Skelly-Sankey Oil Company. In 1937, Getty Oil Company acquired controlling interest in Skelly Oil Company and it became a part of the Getty Oil complex.[23]

Skelly's policy was simple. "There's no room in this industry for just another oil company....But there is room for a company that...[makes] everything it sells *better*....[and] that will be...[our] policy" he declared. Skelly held to a rigid personal routine. He arrived at his office each morning at about 6:00 a.m. and conducted business on the telephone until about 10:30 a.m., when his chauffeur drove him to a nearby barber shop for a shave and a shoeshine. Within 30 minutes, he would be back in his office where he would work the remainder of the day.[24]

Skelly's home, at Twenty-first and South Madison in Tulsa, was not as ostentatious as those of some of the other local oilmen. Nonetheless it cost $250,000 and contained 25 rooms, 8 bedrooms, 5 baths, several half baths, and five fireplaces. Well-known for his lawn parties, which featured mint juleps, Skelly often entertained such notables as Wiley Post, Will Rogers, and Amelia Earheart.[25]

Skelly's greatest contribution to Tulsa and the Oklahoma petroleum industry, however, was his work in the establishment of the International Petroleum Exposition and Congress, later simply the International Petroleum Exposition or IPE. The movement was started on March 3, 1923, by Earl Sneed who, in a story that appeared in *American Saturday Night*, called for Tulsa to host an International Petroleum Congress. Such an affair, Sneed pointed out, would "give thrills to the young people, knowledge to the oil fraternity, opportunity to make world-wide acquaintances, renew friendships, and firmly establish Tulsa for all time to come as the oil center of the entire world."[26]

Quickly grasping the idea, the Tulsa Chamber of Commerce and its Convention Committee soon incorporated the International Petroleum Exposition and Congress, with Edward F. McIntyre as General Manager. The first President of the IPE was L. B. Jackson. Other officers were J. M. Hayner and W. A. Vanderver, Vice President; J. J. McGraw, Treasurer; and William Holden, Secretary. Additional members of the Board of Directors were H. O. McClure, A. V. Bourque, Alf G. Haggem, W. A. Melton, J. H. Gardner, and T. J. Hartman.[27]

For the first IPE 12 working committees were organized to oversee the event—finance, exhibition, attractions, exposition, convention, parades and pageants, transportation, scientific and technical exhibits, public safety, auditing, entertainment, decorations, and buildings and grounds. In addition to the officers and directors of the IPE, numerous Tulsa area oilmen and civic leaders worked to make it a success. Because of their ef-

Opening day ceremonies at one of the early International Petroleum Expositions held in Tulsa. Attended by oilmen from around the world, the IPE did more than anything to make Tulsa the oil capital of the world. (Beryl D. Ford).

forts, within six months of Sneed's article the First IPE opened on October 7, 1923. Built around a carnival atmosphere, it boasted $1,500,000 worth of exhibits that attracted oilmen from around the world. An unqualified success, the Exposition quickly became an integral part of the international petroleum community.[28]

In 1925, W. G. Skelly became president of the Third IPE and remained in office through the Thirteenth International Petroleum Exposition held in 1953. During his tenure, he guided the Exposition through some of its most troubled times, and when the tremendous expense of the IPE forced a rescheduling of the event, Skelly remained adamant that the Exposition be continued. Without his strong personal leadership, the IPE likely would not have survived the Great Depression or World War II.[29]

In 1925 it was decided to hold the IPE bi-annually. However, this decision held only through the 1927 gathering, when the schedule returned to an annual basis through 1930. Because of the effects of the Great Depression, another show was not held until 1934. The gathering returned to a bi-annual schedule for 1936, 1938, and 1940. No shows were held during World War II, but when that conflict ended, Skelly reorganized the meetings and opened the Silver Anniversary IPE on May 15, 1948.[30]

Because the bi-annual schedule did not allow sufficient time between technological innovations to justify its continuation, Skelly and the IPE's other directors in 1948 decided to shift to a five-year program. As a result, the next Exposition was not held until 1953. It proved to be Skelly's last, for he died on April 11, 1957.[31]

Skelly was succeeded as President of the International Petroleum Exposition by William K. Warren, another outstanding Tulsa oilman and philanthropist. Additional shows were held in

W. G. Skelly, longtime president of the International Petroleum Exposition, addressing a crowd on opening day of the 1948 IPE. (PennWell Publishing Company).

1959, 1966, 1971, and 1976. The 1976 gathering was the largest of them all. Coinciding with the American Bicentential Celebration, it featured $2 billion worth of exhibits. Unfortunately, in the late 1970s it was decided to cancel plans for the 1980 IPE. Eventually, because of adverse conditions in the oil and gas industry, this decision was expanded to all future Expositions and the corporation overseeing the IPE was disbanded.[32]

Throughout its life, the International Petroleum Exposition did much to substantiate Tulsa's claim as "Oil Capital of the World." Millions of people from every major oil producing country in the world attended the event, and billions of dollars worth of equipment was displayed. It not only provided a forum for oilmen to express their opinions on governmental policies affecting the industry, but also offered a method for them to keep abreast of the latest technological innovations.[33]

From the first IPE, a diligent effort was made to enhance the diffusion of oil related technology on display. A "special technical department" at Tulsa's Carnegie Library was created in 1923 to serve as a technological archive. For decades the collection was recognized as one of "the world's largest repository of information on the petroleum industry" and was utilized by numerous scientists and researchers from throughout the oil business.[34]

Tulsa's prominence as a technological center of the petroleum industry, started by the IPE's work with the Carnegie Library, was given a boost in 1928 when the University of Tulsa founded its College of Petroleum Engineering. The university had evolved through a series of schools. Founded in Muskogee as the Presbyterian School for Indian Girls in 1882, it became Henry Kendall College in 1894. When the Tulsa Commercial Club promised to provide a 20-acre campus, a streetcar line connecting the school with downtown Tulsa, natural gas to heat and light the buildings for 6¢ per 1,000 cubic feet, and $100,000 cash, the school moved to Tulsa in 1907. It became the University of Tulsa in 1920.[35]

As an almost natural evolution for a school located in the "Oil Capital of the World," the University of Tulsa's School of Petroleum Geology opened in 1928. Owing its existence to the community's petroleum industry, it actually started

A night view of the 1949 IPE. Note the huge American flag, which became a well-known trademake of the Exposition. (Oklahoma Historical Society).

holding classes in the laboratory of the Carter Oil Company. Robert C. Beckstrom of the Colorado School of Mines served as the school's first dean. By 1931, Beckstrom had been joined on the facility by Sidney Born and Wilbur L. Nelson. The three men, with the avid support of Tulsa's oilmen, quickly transformed the school into the "greatest training" facility in refinery production in the world.[36]

In recognition of the support of local oilmen, many of school's buildings bear their names. Kemp Hall was named for E. Roger Kemp; Harwell Gymnasium for E. W. Harwell; the McFarlin Library for Robert M. McFarlin; the Phillips Engineering Building for Waite Phillips; LaFortune Hall for Joseph A. LaFortune; Skelly Stadium for William G. Skelly; Tyrrell Hall for Harry C. Tyrrell; Oliphant Hall for Mr. and Mrs. Allen G. Oliphant; and the Chapman Building for James A. Chapman. Other facilities honoring Tulsa's petroleum leaders are the John Mabee Hall for Men, the Lotta Jane Mabee Hall for Women, the Mabee Gymnasium, the Mabee Gymnasium Annex, Mark K. Chapman Center for Communicative Disorders, and the John Rogers' College of Law Building. Eventually, the courses in the School of Petroleum Engineering were expanded to include petroleum engineering, geological engineering, petroleum geophysics, petroleum land management, and petroleum sciences. Known world-wide, the University of Tulsa attracted thousands of future oilmen to the community for the completion their education.[37]

In conjunction with its growth as a center of petroleum research and education, Tulsa became the home of several petroleum related professional organizations. Tulsan J. Elmer Thomas, during the latter part of 1915, joined with Charles H. Taylor and E. L. DeGolyer, both connected with the University of Oklahoma's Geological Depart-

The Petroleum Engineering Building at the University of Tulsa. The early decision by University of Tulsa leaders to make the institution a leader in oil and gas research helped Tulsa earn the title of oil capital of the world. (University of Tulsa, McFarlin Library).

ment, to form a professional geological organization. Through their efforts an organizational meeting was hosted by Taylor on January 7-8, 1916. The following year, on February 10, 1917, a second gathering was held and the group voted to move to Tulsa and form the Southwestern Association of Petroleum Geologists. In 1918 the group's named was changed to the American Association of Petroleum Geologists, and Tulsa was made its official headquarters. In addition to the AAPG, Tulsa became the home for the the Society of Exploration Geophysicists, and the Oklahoma-Kansas Division of the Mid-Continent Oil and Gas Association.[38]

Unfortunately Tulsa's claim as the "Oil Capital of the World" began to wane during the 1950s, as more and more American companies expanded their overseas operations. Nonetheless, in 1979 Tulsa still boasted 850 oil or petroleum related industries and more than 30,000 petroleum industry workers. In addition, it was the home of numerous regional and international oil industry companies. While Tulsa's recognition as the "Oil Capital of the World" and its continued role as a major center of the energy industry was one aspect of Glenn Pool's legacy, there were numerous others.[39]

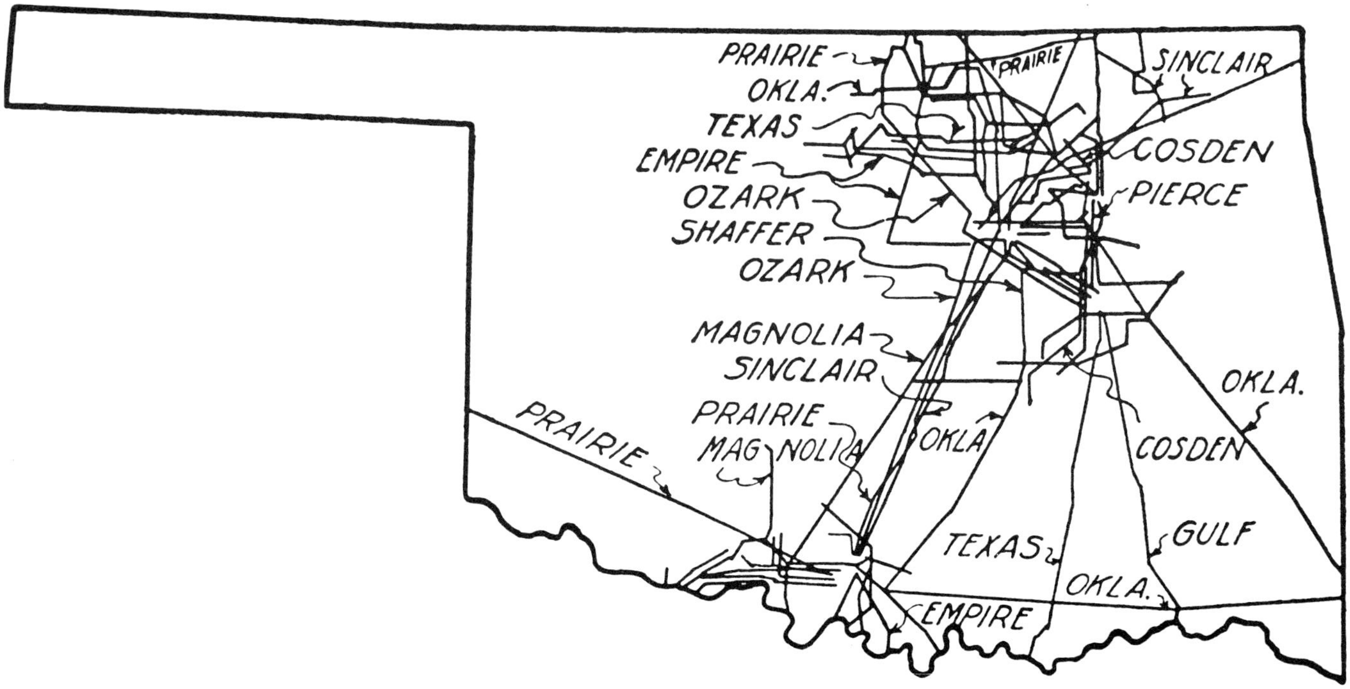

This map of Oklahoma pipelines clearly shows the concentration of lines in the Tulsa area. As the center of such a pipeline network, Tulsa dominated the surrounding communities that once had challenged for superiority in the area. Eventually, Red Fork was annexed to Tulsa and Glenpool became a small "bedroom" suburb. Many of the pipeline companies, such as Prairie, Texas, Sinclair, Cosden, and others, maintained major offices in Tulsa. (Oklahoma Geological Survey).

Robert M. McFarlin, an early petroleum benefactor of the University of Tulsa. (Oklahoma Heritage Association).

The McFarlin Memorial Auditorium on the campus of the University of Tulsa. Because of the support of many local oilmen, the University of Tulsa became a major source of training for petroleum-related professions. (Oklahoma Heritage Association).

James A. Chapman, on the right, was another pioneering Tulsa oilman and supporter of the University of Tulsa. (Oklahoma Heritage Association).

An indoor drilling rig used for training petroleum students at the Unviversity of Tulsa. Such facilities attracted students from around the world to the oil capital. (University of Tulsa, McFarlin Library).

The emblem of the Tulsa Petroleum Club proudly proclaimed the community as the oil capital of the world. (Beryl D. Ford).

CHAPTER ELEVEN

GLENN POOL'S LEGACY

The tremendous production of Glenn Pool overshadowed any other find up to that time in present-day Oklahoma. Total pre-World War I production from the field was an astounding 121,228,301 barrels of crude. Glenn Pool's production, when combined with the discovery of several new fields in California, created a revolution in the American petroleum industry and launched Oklahoma's petroleum boom, a boom that brought billions of dollars into the state during the following decades. It also awakened oilmen to the tremendous possibilities of the great Mid-Continent Region, of which Glenn Pool was a vital part.[1]

Stretching north from Texas, across Oklahoma, into Kansas, and west from New Mexico to Louisiana, the Mid-Continent, within three decades of Glenn Pool's discovery, became one of America's greatest oil and gas producing areas. Between 1900 and 1935, 8,804,000,000 barrels of crude were extracted from the numerous pools of the Mid-Continent. In 27 of those 35 years, the region ranked first among the nation's oil producing areas, and between 1918 and 1922 and 1924 and 1935, the Mid-Continent Region accounted for more than half of all the oil produced in the United States. Much of this early production came from Glenn Pool's wells, which in 1909 and 1910 accounted for approximately 40 percent of the Mid-Continent's output.[2]

Glenn Pool's discovery also had a tremendous impact on Oklahoma's infant petroleum industry. By 1905 the state already ranked eighth among the nation's oil producing states; however, because of Glenn Pool's production, by 1906, Oklahoma had jumped into second place, and at statehood in 1907 it leaped into first place. Oklahoma maintained its lead in 1908 before dropping to second place in 1909, 1910, 1911, 1912, 1913, and 1914 and then regaining the lead in 1915, 1916, and 1917.[3]

Glenn Pool proved to be one of America's most prolific sources of crude. By the close of 1949 the field's cummulative production stood at 238,535,000 barrels of oil, sufficient to make Glenn Pool the 20th most productive oil field ever uncovered in the United States. Total output during the first 75 years of Glenn Pool's life amounted to more than 300 million barrels of oil.[4]

CRUDE PRODUCTION, 1907-1914[5]
(42 gallon barrels)

	Glenn Pool	Mid-Continent	United States
1907	19,927,000	46,896,000	166,095,000
1908	20,494,000	48,824,000	178,527,000
1909	18,947,000	50,834,000	183,171,000
1910	19,237,000	59,218,000	209,557,000
1911	13,880,000	66,595,000	220,449,000
1912	10,496,000	65,473,000	222,935,000
1913	9,470,000	84,920,000	248,446,000
1914	8,678,000	97,995,000	265,763,000
1915	5,994,000	123,294,000	281,104,000
1916	7,282,000	136,934,000	300,767,000

While Glenn Pool's tremendous outpouring insured its place in the petroleum heritage of America, its discovery had two additional important effects upon the nation's petroleum industry. The completion of the Ida Glenn No. 1 came at a time when the production of the Texas-Louisiana Gulf Coast fields was dropping. At the same time, Glenn Pool's crude, which was "rich in gasoline content, easy to refine, and equally suitable for fuel oil"—three assets which made it attractive to large refining concerns—offered a viable alternative source of supply. The timing of its discovery and the value of its crude enabled Glenn Pool to rescue the floundering Gluf Coast refining business and insure the region's importance to the American petroleum industry.[6]

For example, The Texas Company (Texaco) had originally been chartered on April 7, 1902, and had its entire future tied to the output of Spindletop. Because at first there seemed to be no end to Spindletop's output, which in 1902 stood at approximately 17.5 million barrels, The Texas Company contracted for 1 million barrels of crude and quickly began to purchase equipment to handle this amount. Fifteen steel tanks were built at Garrison, Texas; 12 at Nederland, Texas; 7 at Port

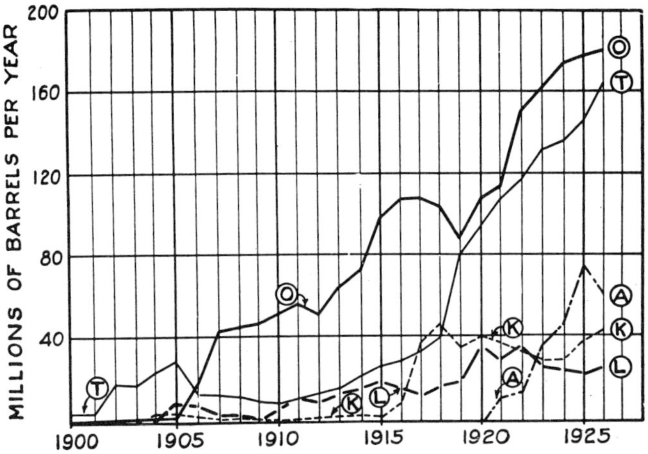

This graph, which illustrates the relationship between the oil producing states of Oklahoma (O), Texas (T), Arkansas (A), Kansas (K), and Louisiana (L), clearly shows the impact of Glenn Pool on the production of the region. During the first 20 years from 1906, the discovery at Glenn Pool, through 1926, Oklahoma led the region in oil production. (Oklahoma Geological Survey).

Arthur; and 2 at Amesville, Louisiana. Another earthen tank holding 103,000 barrels was constructed, and 10 railroad tank cars and the 400-foot *Texas Barge No. 1* were purchased. Work was started on a $150,000 refinery and $20,000 asphalt plant at Port Arthur to process the crude. Within 8 months, The Texas Company had spent $652,040.72 to refine and market Spindletop oil.[7]

To justify such an expenditure, The Texas Company by 1902 had developed a sophisticated marketing chain. Spindletop crude was purchased and either resold to northern refineries or refined and sold to sugar plantations in Louisiana as fuel for the sugar cane harvest. Another agreement with the Producers Oil Company called for The Texas Company to supply 1 million barrels of oil to Producers at 25¢ per barrel.[8]

However, "all the ambitious plans of The Texas Company depended on a plentiful supply of oil from the Southeast Texas wells, which, in 1902, neant Spindletop. Suddenly, "the hopes based on these moves collapsed like a punctured balloon" when in the fall of 1902 Spindletop's wells "ceased to flow." Production in the field dropped from 62,000 barrels daily to 20,000, and within another year had fallen to 5,000 barrels. "Without fresh sources of oil, The Texas Company...would be done for."[9]

Although other sources were found at Sour Lake, Batson, and Saratoga, Texas and Jennings, Louisiana, the replacement oil, especially that from Sour Lake and Jennings, was not as high a grade as the officials of The Texas Company wanted. While it was adequate for "gas oil," which was used to "enrich artificial gas used for lighting and cooking," it did not refine into high-grade kerosene or gasoline. While the market for gasoline at this time was rather limited—the automobile age had not yet dawned—kerosene was in great demand.[10].

To further complicate The Texas Company's problem, just when the use of kerosene declined because of the availability of electricity, there was a dramatic surge in the demand for gasoline. For example, the domestic use of gasoline grew from approximately 5 million barrels in 1904 to nearly 11.3 million barrels in 1909. Correspondingly, the number of automobiles in America grew from approximately 23,000 in 1902 to 902,000 in 1912. In the same time span, the number of trucks increased from practically none to 41,000. By 1909, the manufacturing of gasoline was "absorbing about 25 percent of total domestic distribution," and in 1912 total production of gasoline reached 20.3 million barrels. Accompanying the growth in demand was a rise in gasoline price, which by 1918 was approximately 24.2¢ per gallon.[11]

For The Texas Company to remain competitive it needed to expand into the gasoline market. To tap into the new market it was necessary to find another source of high grade oil. Obviously the closest and most easily accessable was Glenn Pool's crude, which produced twice as much kerosene and five times as much gasoline as the oil available in East Texas and Louisiana.[12]

Joseph S. Cullinan, president of The Texas Company, "quickly grasped the significance of Glenn Pool." The major drawback, however, was that the cost of transporting the crude by railroad was more than the price of the oil. To solve the problem, Cullinan decided to extend the existing Port Arthur-to-Humble, Texas pipeline northward to Tulsa. The decision was to prove "incalculable" on the future history of Texaco. The proposed line was 473 miles long, and cost $5,641,000 dollars, but by November of 1907 it

The Texas Company's pump station in Glenn Pool. In February, 1907, The Texas Company began work on a pipeline connecting Glenn Pool with its Port Arthur, Texas, refinery. Costing $5,641,000 the line was completed in November, 1907, and insured the future of the company. (Oklahoma Historical Society).

The Gulf Oil Company's pump station in Kiefer. In February, 1907, the Gulf Production Company started work on another 8-inch pipeline that stretched 413 miles from Glenn Pool to Sour Lake, Texas. It was opened in August, 1907, and offered another outlet for Glenn Pool crude. (Sapulpa Historical Museum).

was carrying Glenn Pool oil south. The pipeline proved to be "a money-maker from the start."[13]

Like The Texas Company, the other large Gulf Coast refinery firm, the Gulf Refining Company of Texas, which later evolved into Gulf Oil Corporation, faced the same problem of securing adequate crude for its Port Arthur refinery complex when Spindletop's production declined. Although temporary replacement crude was located in Texas and Louisiana, in the fall of 1906 a report from a Gulf Stockholders Committee, consisting of A. W. Mellon, William Flinn, Emmet Queen, James D. Callery, and J. H. Reed, stated that in the entire state of Texas there was sufficient production of high grade crude only to supply 12,000 barrels daily to all refineries. Mellon and the others also pointed out that the total capacity of the Gulf Refining Company equaled this amount and thus "if the company could get all the refinable crude produced in Texas, it would hardly be enough." Thus, Mellon concluded, if Gulf planned to meet its competition it would be "necessary for the Company to enter the Indian Territory Field [Glenn Pool]" and tap its high-grade oil[14]

Work was under way by November, 1906, to unite the Gulf Company's refineries and the Glenn Pool wells. Working from "daylight to dark, 3,000 men swung picks and shovels digging a 400-mile ditch four feet deep." By August of 1907 the line was completed and shortly thereafter Glenn Pool crude began flowing to Gulf's Port Arthur refinery.[15]

While Glenn Pool's discovery secured the future of such companies as Texaco and Gulf, their pipelines into the field insured the future of the Oklahoma petroleum industry. Although there had been several discoveries in the Twin Territories prior to Glenn Pool, oilmen always had been plagued by the question of how to market their product. To have oil is one thing, to sell it for a profit is another.[16]

For example, after Cudahy had completed the Nellie Johnstone No. 1 near Bartlesville in 1897 with "a good-sized stream [of oil] along the ground," he discovered that because there were no nearby railroads or pipelines the only means of marketing the crude was to sell it to local farmers for illumination, medicinal purposes, and livestock dip. Obviously such a limited market quickly was swamped with oil. When it became clear that he would not be able to sell the well's output, Cudahy was forced to cap the Nellie Johnstone.[17]

While some of Oklahoma's early crude production could be shipped to market by railroad, the cost of shipping the oil usually was prohibitive. For example, it cost 78¢ per barrel to ship Glenn Pool crude by rail to Port Arthur or 90¢ per barrel to send the oil to Neodesha, Kansas, the two closest refineries. Because of the limited local market and the high cost of shipping crude by rail, the development of Oklahoma's petroleum deposits had lagged seriously prior to the Ida Glenn No. 1.[18]

Before the discovery at Glenn Pool, there had been only one major pipeline completed into the region, when, in 1904, the Prairie Oil and Gas Company constructed a 6-inch pipeline from Independence, Kansas, to Bartlesville and Ramona in the Cherokee Nation. From Independence, the line ran northeast through Kansas City, Missouri, to the huge refinery complex at Whiting, Indiana. From there it was piped to outlets on the Atlantic Coast. Although this line eventually was built southward to Red Fork and then, in February, 1906, tied to Glenn Pool by a 2-inch extension, it obviously was inadequate to handle the field's output by that time.[19]

With the completion of The Texas Company's and Gulf's pipelines connecting the Gulf Coast and Glenn Pool, Oklahoma's petroleum industry gained a vast new market for its crude. The only reason for the construction of these lines was Glenn Pool. If the field had not been uncovered, the state's oil business would have continued to lag, with no adequate method of marketing its vast quantities of crude.[20]

Once tied to the Gulf Coast, Oklahoma's pipeline network grew tremendously. By 1926, the Prairie, Texas, and Gulf company pipelines had been joined by those of the Cosden, Sinclair, Pierce, Oklahoma, Empire, Ozark, Shaffer, and Magnolia companies to form a 19,000-mile-network converging on the Glenn Pool-Tulsa area. The availability of adequate pipeline facilities likewise encouraged the search for more oil.[21]

Adequate markets and the availability of high-grade Glenn Pool crude helped create another new petroleum industry in the state—the manufacturing of gasoline. Glenn Pool's crude, like most of Oklahoma's oil, produced a large amount of gasoline after it was refined, with the average bar-

The loading rack of the D. W. Franchot Company's natural gasoline plant at Kiefer. The facility was the first of its kind in the state and was made possible by Glenn Pool's high-grade crude. (Sapulpa Historical Museum).

Following Franchot's initial experiment with casinghead gasoline at Kiefer, several other plants quickly were opened. This is the Gypsy Oil Company's casinghead gasoline plant at Kiefer. The natural gasoline industry grew so quickly in the state after the discovery at Glenn Pool that by 1913 Oklahoma was producing one-fourth of the nation's supply of natural gasoline. (PennWell Publishing Company).

rel of Oklahoma oil yielding 50 percent gasoline, 8.5 percent kerosene, 33 percent fuel oil, 2.5 percent lubricating oil, and 6 percent various other petroleum products and losses in the refining process. Much of Oklahoma's early gasoline production came from casinghead gas or natural gasoline. Consisting of liquids found in "wet natural gas," it initially was converted into "natural gasoline" by normal condensing that occurred in gas pipelines.[22]

However, as gasoline became an increasingly greater percentage of the crude oil market, the casinghead gas was converted into gasoline in special compression, absorption or refrigeration refining facilities. The first of these was opened in Oklahoma by the D. W. Franchot Company near Kiefer in 1909. Its initial output was approximately 24 barrels daily.[23]

Afterward Oklahoma's natural gasoline industry quickly grew. Within a decade of the opening of Franchot's plant, the state had become a major supplier of natural gasoline, and by 1913, Oklahoma was producing one-quarter of America's overall output of natural gasoline. The amount of production continued to rise until by 1915, the state's output of natural gasoline was in excess of one-half of America's total production.[24]

NATURAL GASOLINE PRODUCTION[25]
(gallons)

	Oklahoma	**Nation**
1909	366,912*	—
1910	—	—
1911	388,058	7,000,000
1912	1,575,644	12,000,000
1913	6,462,968	24,000,000
1914	17,277,555	43,000,000
1915	31,665,991	65,000,000
1916	48,359,602	103,000,000
1917	115,123,424	218,000,000
1918	163,700,550	283,000,000
1919	189,995,038	352,000,000

(*approximately)

By locating their plants in the Glenn Pool area, refineries streamlined their operation by being close to their source. The construction of the Tulsa-Glenn Pool refinery complex was aided by the fact that Tulsa businessmen encouraged the location of refineries in the community. As a result, the region became a major base for both compression refineries, which reduced the natural gas to liquid fractions by compression, and cooling or absorption refineries, which recovered natural gasoline from "wet" gas. By 1926 there were 22 refineries located in the area.[26]

TULSA-GLENN POOL REFINERIES[27]

Company	**Type**	**Daily Capacity in gallons**
A.C.F. Gasoline Co. (Tulsa)	Compression	750
Amity Gasoline Co. (Mounds)	Compression	300
Arthur Oil Co. (Sapulpa)	Compression	1,000
Aztec Oil Co. (Jenks)	Compression	1,500
Barnsdall Oil Co. (Tulsa)	Absorption	2,500
Chestnut & Smith Corp. (Kiefer)	Compression and Absorption	20,500
Crosbie & Gillespie (Kiefer)	Compression	7,500
Diamond Refining Co. (Jenks)	Absorption	600
Eagle Gasoline Co. (Kiefer)	Compression	3,000
Four Gasoline Co. (Tulsa)	Compression	500
Glenn Gas Co. (Glenn Pool)	Compression	2,000
Gypsy Oil Co. (Kiefer)	Compression	48,100
Gypsy Oil Co. (Jenks)	Compression	2,200
National Production Co. (Mounds)	Absorption	500
Oil State Gasoline Co. (Jenks)	Compression	1,500
Oklahoma Gasoline Plants, Inc. (Sapulpa)	Absorption	5,500
Sand Springs Home (Sand Springs)	Compression Absorption	3,000
Southland Gasoline Co. (Sapulpa)	Compression	1,500
The Texas Company (Jenks)	Compression	804
Transcontinental Oil Co. (Tulsa)	Absorption	9,000

The Oklahoma Natural Gas Company plant at Sapulpa. By 1907, the fledgling Oklahoma Natural Gas Company had connected Oklahoma City to Glenn Pool's gas wells. This was the first attempt in the state to market natural gas on a large scale and marked the beginning of Oklahoma's natural gas industry. (Oklahoma Historical Society).

Company	Type	Daily Capacity in gallons
Valley Osage Oil Co. (Sapulpa)	Absorption	2,000
J. H. Wright (Sapulpa)	Absorption	4,000

Glenn Pool also proved to be the major source for the state's young natural gas industry. By early 1907, the field reported 12 gas wells. Although much of the natural gas was simply burned off or flared into the atmosphere, as the natural gasoline market grew so did conservation efforts.[28]

At first, the lack of adequate markets made it financially difficult to develop an independent gas industry. However, there were several early efforts. In 1903, James M. Guffey and John Galey drilled several wells in what became the Greater Osage Field. After encountering natural gas, the two oilmen piped it through a 2- and 3-inch pipeline system to Tulsa, where it supplied fuel for a brick plant. Later they began selling the gas to individual customers for lighting and heating, and by 1905 they were serving 400 households in the region. To operate their system, Guffey and Galey organized the Arkansas Valley Oil and Gas Company.[29]

Later in 1905, two Pittsburgh, Pennsylvania oilmen, Theodore N. Barnsdall and Glenn T. Branden, purchased the Arkansas Valley Oil and Gas Company, minus the Tulsa service, which they operated as the Osage and Oklahoma Company. Barnsdall became so involved in the natural gas industry that by the following year, 1906, he also had acquired the Tulsa distribution system. That same year, Robert Galbreath formed the Gal-

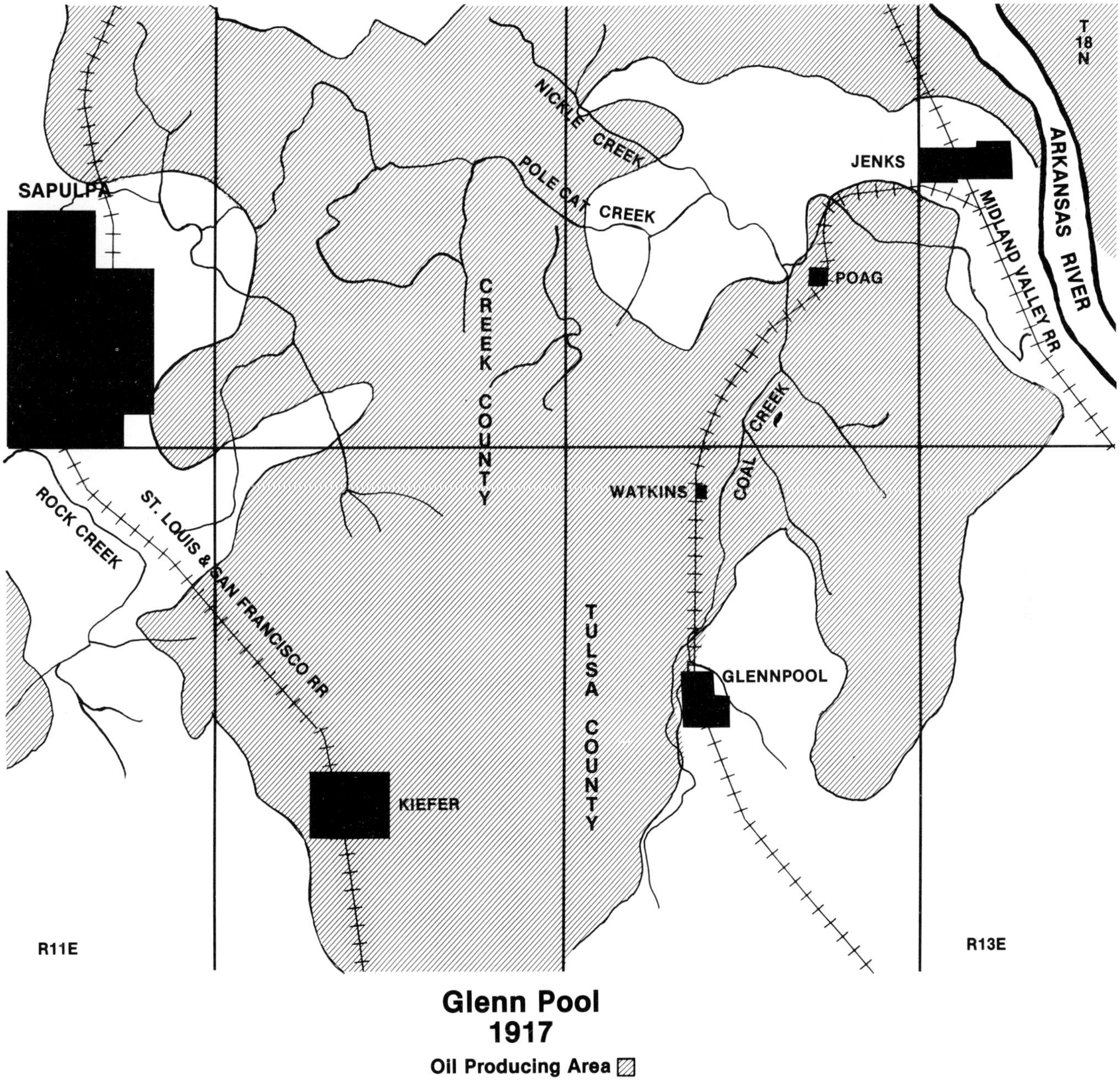

Glenn Pool, 1917. (Oklahoma Heritage Association).

breath Gas Company, constructed a 6-inch pipeline into Tulsa, and began marketing Glenn Pool natural gas in competition with Barnsdall. Their rivalry quickly drove down the price of natural gas to 5¢ per 1,000 cubic feet.[30]

Galbreath's and Barnsdall's price war ended expansion into the Tulsa market for the time being, but several other pioneering natural gas entrepreneurs wondered if Glenn Pool's natural gas could not be used to light and heat communities elsewhere in the state. Although many potential financial backers questioned whether such a business could survive in what they assumed "to be a hot and dry country," Dennis T. Flynn and C. B. Ames were convinced that Glenn Pool's natural gas could be transported to central Oklahoma and sold competively with the Oklahoma Gas and Electric Company's manufactured gas. Because they were attorneys for the H. M. Byllesby Company of Chicago, Flynn and Ames approached Byllesby with their proposal.[31]

Impressed with the concept, Byllesby on October 9, 1906, agreed to join them in forming the Oklahoma Natural Gas Company (ONG). Originally their plan was to connect Oklahoma City with the Cleveland Field; however, when it became apparent that the Cleveland wells were being depleted rapidly, Byllesby withdrew his support. After selling Byllesby's interest to Glenn T. Braden and Theodore N. Barnsdall, Flynn and Ames started again. In the reorganization, Flynn was elected President and Ames Assistant Secretary, with R. H. Bartlett, an associate of Braden and Barnsdall, representing their interest on ONG's board of directors.[32]

Abandoning the idea of tapping Cleveland's wells, Flynn and the others began building northeast from Oklahoma City toward Glenn Pool in the spring of 1907. By the fall of that year, the line was completed at a total cost of $1.7 million. Oklahoma City received its first shipment of Glenn Pool natural gas on December 28, 1907.[33]

The initial attempt to tap the Oklahoma City market proved so successful that two 8-inch branch lines were extended to Guthrie and Shawnee. To supply the new customers several additional wells were drilled in Glenn Pool near Mounds in 1909 by ONG. The following year, 1910, a 10-inch pipeline was constructed to carry this gas to Kellyville, where, that same year, Oklahoma's first compressor station was built.[34]

ONG's effort at Glenn Pool was the first real attempt to market natural gas in the state and marked the beginning of one of the state's most important businesses—the natural gas industry. During Glenn Pool's heyday, Oklahoma's production of natural gas jumped from 3,520,396,000 cubic feet in 1906 to 87,516,753,000 in 1915. By that time Oklahoma was producing approximately 12 percent of America's total output of marketed natural gas.[35]

Thomas Gilcrease, whose Glenn Pool leases enabled him to assemble one of the finest private collection of Americana in the United States. (Oklahoma Heritage Association).

Alfred E. Aaronson, the man who led the movement to keep Thomas Gilcrease's collection of art in Tulsa. (Oklahoma Heritage Association).

The Thomas Gilcrease Institute of American History and Art. Originally the private collection of Thomas Gilcrease, an early-day Glenn Pool oilman, the Institute became a private foundation, open to the public, with hundreds of millions of dollars worth of art and artifacts on display. (Oklahoma Historical Society).

MARKETED NATURAL GAS PRODUCTION[36]

	Oklahoma	**Nation**
1906	3,520,396,000	388,843,000,000
1907	4,867,031,000	406,622,000,000
1908	11,924,574,000	402,141,000,000
1909	47,859,218,000	480,706,000,000
1910	52,028,718,000	509,155,000,000
1911	56,069,608,000	512,993,000,000
1912	73,799,319,000	562,203,000,000
1913	75,017,688,000	581,898,000,000
1914	78,167,414,000	591,867,000,000
1915	87,516,753,000	628,579,000,000

Glenn Pool's outpouring of quality oil and natural gas provided the catalyst needed to ignite the Oklahoma oil rush, and with the boom came the rise of numerous oilmen/philanthropists within the state who left enduring legacies for all Oklahomans. One of the greatest legacies growing out of Glenn Pool was that left by Thomas Gilcrease and Alfred E. Aaronson.[37]

The son of a Scotch-Irish father and Creek Indian mother, Gilcrease was born in Louisiana on February 8, 1890. While he was still a child, his family moved to the Creek Nation in Oklahoma, and, because of his mother's headright, Gilcrease received a 160-acre allotment when he was 14 years old. Located about two and a half miles east of Keifer and a mile south of the Arkansas River, the property was approximately 15 miles south of Tulsa and squarely in the middle of the Glenn Pool. Within a short time after the strike, Gilcrease's property was producing nearly a million dollars worth of crude annually.[38]

Initially, Gilcrease had difficulty in securing control over the development of his allotment. It originally had been leased to W. H. Milliken, who in return for paying Gilcrease $90,000, annually retained another $900,000 per year. Realizing that he could be receiving more if he could gain control of the lease, Gilcrease took Milliken to court. Although the legal action lasted nearly a decade, Gilcrease won. By 1917, his allotment contained 32 producing wells, and Gilcrease parlayed his holdings into a vast oil empire organized under the Gilcrease Oil Company.[39]

Gilcrease's first love was art. He had purchased his initial painting, "Rural Courtship," in 1912, and over the years acquired one of the largest private collections of Western American Art and archival material in the world. So that his collection could be displayed properly to the public, Gilcrease started construction, in 1949, of a museum next to his home on West Newton in Tulsa. However, before his dream was realized, he suffered a series of business reverses that by 1954 placed him in serious financial trouble. To satisfy his creditors, Gilcrease faced the prospect of selling his paintings, which, of course, would deprive Tulsa of a major art collection.[40]

Fortunately, Alfred E. Aaronson became involved in the problem at this time. Born in New York City in 1893, Aaronson had migrated to Oklahoma in 1913, and entered the petroleum industry. Serving as Vice-President of the Mid-Continent Gas Company from 1913 to 1918; the Tuloma Oil Company from 1915 to 1926; and Commonwealth Company from 1925 to 1935,

Aaronson became a familiar figure among Tulsa's oil fraternity.[41]

As Aaronson recalled, after learning that Gilcrease's collection would probably be leaving Tulsa, he and his wife decided to "take one last look at it." Although he had never previously met Gilcrease, while at the museum "I went over to talk to him and to ask whether it was true that he was about to move the collection to some other area." "Yes, the creditors are moving in," Gilcrease replied. On the spot, Aaronson declared, "I'm not a rich man, but I'd give $1,000 to keep that thing here." This was the beginning of the drive to save one of Tulsa's most valuable art treasures.[42]

As Aaronson and his wife walked through the galleries, he later recalled, "I met several of my friends and told them what I'd said." "They said that they would give some money to keep it in Tulsa, too," he continued. Discussing while they drove home what a loss the sale of Gilcrease's paintings would mean to Tulsa, Aaronson and his wife decided to attempt to raise sufficient funds to purchase the collection.[43]

After agreeing to pay for the election, Aaronson convinced L. C. Clark, the Mayor of Tulsa, to hold a community-wide referendum on the question of the city acquiring the Gilcrease Collection. By a three to one margin, the people of Tulsa voted to take custody of the numerous art works, Gilcrease's home, the museum building, and 13 acres of land. The sale was completed in July of 1955, and shortly afterward the Thomas Gilcrease Institute of American History and Art opened. Between $65 million and $100 million worth of "fabulous and irreplaceable... paintings, etchings, folios, sketches, books, sculptures in bronze, wood and stone, pre-Columbian treasures and artifacts of Americana," were available for public viewing.[44]

The Gilcrease Collection is but one example of how Glenn Pool's legacy outlived the wells. In addition to making Tulsa the undisputed center of the oil industry for several decades, Glenn Pool gave birth to several world-wide energy concerns and launched numerous oil fortunes. However, more than anything, Glenn Pool freed Oklahoma oilmen of the necessity to rely on Eastern capital to develop the state's petroleum reserves. With the huge outpouring of crude from Glenn Pool's wells, state oilmen now had the financial capital to continue the search for crude that eventually led to a series of prolific finds, including Cushing, Healdton, the Osage, Seminole, and Oklahoma City.[45]

ENDNOTES

CHAPTER ONE
Red Fork

[1]Kenny A. Franks, Paul F. Lambert, and Carl N. Tyson, *Early Oklahoma Oil: A Photographic History, 1859-1936* (College Station: Texas A&M University Press, 1981), p. 108; Robert Gregory, *Oil in Oklahoma* (Tulsa: Leake Industries, Inc., 1976), pp. 7, 14.

[2]Glen S. Dille, "Economic Geology: Oil and Gas in Tulsa County,"in Malcolm C. Oakes, ed., *Geological and Mineral Resources of Tulsa County Oklahoma, Bulletin No. 69* (Norman: Oklahoma Geological Survey, 1952), p.122; "Tulsa County Map,"Oklahoma Department of Transportation, Oklahoma City, Oklahoma; Bess Mills-Bullard, comp., "Digest of Oklahoma Oil and Gas Fields," *Oil and Gas in Oklahoma, Oklahoma Geological Survey Bulletin No. 40,* Vol. I (July, 1928), p. 230.

[3]Dille, "Economic Geology," p. 122; "Tulsa County Map"; Mills-Bullard, comp., "Digest of Oklahoma Oil and Gas Fields," p. 230; Glen R. Roberson, *City of the Plains: The History of Tulsa Oklahoma* (Ph.D. Dissertation, Oklahoma State University, 1977), pp. 1-9; George H. Shirk, *Oklahoma Place Names* (Norman: University of Oklahoma Press, 1966), p. 209.

[4]Roberson, *City of the Plains*, pp. 9-11; Shirk, *Oklahoma Place Names*, p. 209.

[5]Roberson, *City of the Plains*, pp. 9-11; Shirk, *Oklahoma Place Names*, p. 209.

[6]Roberson, *City of the Plains*, pp. 11-29; John W. Morris, Charles R. Goins, and Edwin C. McReynolds, *Historical Atlas of Oklahoma* (Norman: University of Oklahoma Press, 1976), map 41.

[7]Gaston Litton, *History of Oklahoma at the Golden Anniversary of Statehood* (4 vols., New York: Lewis Historical Publishing Company, 1957), Vol. II, pp. 112-113, 121; Malcolm C. Oakes, "Geography,"in Malcolm C. Oakes, ed., *Geological and Mineral Resources of Tulsa County Oklahoma, Bulletin No. 69* (Norman: Okahoma Geological Survey, 1952), p. 15.

[8]Fred S. Clinton, "First Oil and Gas Well in Tulsa County," *The Chronicles of Oklahoma*, Vol. XXX, No. 3 (Autumn, 1952), p. 326.

[9]*Ibid.*, pp. 326-329.

[10]*Ibid.*, p. 327; Lalia Phipps Boone, *The Petroleum Dictionary* (Norman: University of Oklahoma Press, 1952), p. 102; Kenny A. Franks, *The Oklahoma Petroleum Industry* (Norman: University of Oklahoma Press, 1980), pp. 3-12, 200.

[11]Franks, Oklahoma Petroleum Industry, pp. 12-14.

[12]Morris, Goins, and McReynolds, *Historical Atlas of Oklahoma*, map 34; Franks, *Oklahoma Petroleum Industry*, pp. 8-9; Charles J. Kappler, comp., *Indian Affairs: Laws and Treaties* (5 vols., New York: AMS Press Inc., 1971), Vol. II, p. 390.

[13]Franks, *Oklahoma Petroleum Industry*, pp. 8-9; Kappler, comp., *Indian Affairs: Laws and Treaties*, Vol. II, p. 390.

[14]Kappler, comp., *Indian Affairs: Laws and Treaties*, Vol. I, pp. 33-36; Franks, *The Oklahoma Petroleum Industry*, pp. 50-51.

[15]Kappler, comp., *Indian Affairs: Laws and Treaties*, Vol. I, pp. 93-94; Franks, *The Oklahoma Petroleum Industry*, pp. 21-22, 50-51.

[16]Kappler, comp., *Indian Affairs: Laws and Treaties*, Vol. I, p. 97; Franks, *The Oklahoma Petroleum Industry*, pp. 21-22, 50-51.

[17]Franks, *The Oklahoma Petroleum Industry*, pp. 8-9, 21-22, 50-51; Kappler, comp., *Indian Affairs: Laws and Treaties*, Vol. I, pp. 93-94.

[18]Kappler, comp., *Indian Affairs: Laws and Treaties*, Vol. I, pp. 739, 761, 765.

[19]Franks, Lambert, and Tyson, *Early Oklahoma Oil*, p. 33; Joe Donald Roberts, "Ten-Barrel 'Whodunit' at Red Fork," *The Chronicles of Oklahoma*, Vol. LVI, No. 4 (Winter, 1978-1979), pp. 442-444; Carl C. Rister, *Oil! Titan of the Southwest* (Norman: University of Oklahoma Press, 1949), pp. 57-60.

[20]Clinton, "First Oil and Gas Well in Tulsa County,"p. 327; Franks, *The Oklahoma Petroleum Industry*, p. 200; Boone, *The Petroleum Dictionary*, p. 102.

[21]Franks, Lambert, and Tyson, *Early Oklahoma Oil*, pp. 16-19.

[22]Fred S. Clinton, "The Beginning of the International Petroleum Exposition and Congress,"*The Chronicles of Oklahoma*, Vol. XXVI, No. 4 (Winter, 1948-1949), p. 479; Franks, *The Oklahoma Petroleum Industry*, p. 27; Clinton, "First Oil and Gas Well in Tulsa County," p. 327.

[23]Clinton, "The Beginning of the International Petroleum Exposition and Congress," p. 479; Franks, *The Oklahoma Petroleum Industry*, p. 27.

[24]Franks, Lambert, and Tyson, *Early Oklahoma Oil*, pp. 107-110; Franks, *The Oklahoma Petroleum Industry*, pp. 34-36, 161; Clinton, "First Oil and Gas Well in Tulsa County," p. 327.

[25]Franks, *The Oklahoma Petroleum Industry*, p. 27.

CHAPTER TWO
The Sue A. Bland No. 1

[1]Roberts, "Ten-Barrel 'Whodunit' at Red Fork," pp. 442-444; Clinton, "The Beginning of the International Petroleum Exposition and Congress," pp. 479-480; Franks, Lambert, and Tyson, *Early Oklahoma Oil*, p. 33.

[2]Franks, *The Oklahoma Petroleum Industry*, p. 28.

[3]Bob Gregory, "How *Glenn Pool* Oil Turned to Gold For Tulsa," *Tulsa Magazine*, Vol. 58, No. 5 (November 20, 1980), p. 9; Franks, Lambert, and Tyson, *Early Oklahoma Oil*, pp. 33-34; Franks, *The Oklahoma Petroleum Industry*, p. 29; Roberts, "Ten-Barrel 'Whodunit' at Red Fork," p. 447; Rister, *Oil!*, pp. 82-82.

[4]Franks, *The Oklahoma Petroleum Industry*, pp. 28-29.

[5]*Ibid.*, p. 29; Roberts, "Ten-Barrel 'Whodunit' at Red Fork," p. 443; Franks, Lambert, and Tyson, *Early Oklahoma Oil*, p. 33.

[6]Franks, Lambert, and Tyson, *Early Oklahoma Oil*, p. 33;

Franks, *The Oklahoma Petroleum Industry*, p. 29; Clinton, "The Beginning of the International Petroleum Exposition and Congress," p. 479; "Tulsa County Map"; "Red Fork-Kiefer Map," *Oklahoma Geological Survey Bulletin No. 19*, Part II (April, 1917), Plate XVII.

[7]Franks, Lambert, and Tyson, *Early Oklahoma Oil*, p. 33; Franks, *The Oklahoma Petroleum Industry*, pp. 27-29; Roberts, "Ten-Barrel 'Whodunit' at Red Fork," pp. 446-447; Rister, *Oil!*, pp. 82-82; Clinton, "First Oil and Gas Well in Tulsa County," pp. 312-313.

[8]Franks, Lambert, and Tyson, *Early Oklahoma Oil*, p. 33; Franks, *The Oklahoma Petroleum Industry*, pp. 27-29; Roberts, "Ten-Barrel 'Whodunit' at Red Fork," pp. 446-447; Rister, *Oil!*, pp. 82-82; Clinton, "First Oil and Gas Well in Tulsa County," pp. 312-313.

[9]Gregory, "How Glenn Pool Oil Turned to Gold For Tulsa," p. 10; Franks, Lambert, and Tyson, *Early Oklahoma Oil*, pp. 33-34; Franks, *The Oklahoma Petroleum Industry*, p. 29; Roberts, "Ten-Barrel 'Whodunit' at Red Fork," p. 447; Rister, *Oil!*, pp. 82-82.

[10]Franks, Lambert, and Tyson, *Early Oklahoma Oil*, pp. 33-34; Franks, *The Oklahoma Petroleum Industry*, p. 29; Rister, *Oil!*, p. 83.

[11]Franks, Lambert, and Tyson, *Early Oklahoma Oil*, pp. 33-34; Franks, *The Oklahoma Petroleum Industry*, p. 29; Rister, *Oil!*, pp. 83-83.

[12]Clinton, "First Oil and Gas Well in Tulsa County," p. 313n; Franks, *The Oklahoma Petroleum Industry*, p. 27.

[13]Clinton, "First Oil and Gas Well in Tulsa County," p. 317n.

[14]*Ibid.*, p. 312; Franks, *The Oklahoma Petroleum Industry*, p. 27.

[15]Clinton, "First Oil and Gas Well in Tulsa County," pp. 312-313; Franks, *The Oklahoma Petroleum Industry*, p. 27; Clinton, "The Beginning of the International Petroleum Exposition and Congress," p. 479.

[16]Clinton, "First Oil and Gas Well in Tulsa County," pp. 312-313; Franks, *The Oklahoma Petroleum Industry*, p. 27.

[17]Franks, *The Oklahoma Petroleum Industry*, p. 27; Clinton, "First Oil and Gas Well in Tulsa County," pp. 313-314.

[18]Franks, *The Oklahoma Petroleum Industry*, pp. 27; Clinton, "First Oil and Gas Well in Tulsa County," pp. 313-314.

[19]Franks, *The Oklahoma Petroleum Industry*, pp. 27; Clinton, "First Oil and Gas Well in Tulsa County," pp. 313-314.

[20]Franks, *The Oklahoma Petroleum Industry*, p. 28; Clinton, "First Oil and Gas Well in Tulsa County," pp. 313-314.

[21]Franks, *The Oklahoma Petroleum Industry*, p. 28; Clinton, "First Oil and Gas Well in Tulsa County," pp. 313-314; Franks, Lambert, and Tyson, *Early Oklahoma Oil*, p. 34.

[22]Franks, *The Oklahoma Petroleum Industry*, p. 28; Clinton, "First Oil and Gas Well in Tulsa County," pp. 313-314, 322; Franks, Lambert, and Tyson, *Early Oklahoma Oil*, p. 34.

[23]Franks, *The Oklahoma Petroleum Industry*, p. 28; Clinton, "First Oil and Gas Well in Tulsa County," pp. 313-314, 322; Franks, Lambert, and Tyson, *Early Oklahoma Oil*, p. 34.

[24]Franks, *The Oklahoma Petroleum Industry*, p. 28; Clinton, "First Oil and Gas Well in Tulsa County," pp. 314-315, 330; Franks, Lambert, and Tyson, *Early Oklahoma Oil*, p. 34.

[25]Franks, *The Oklahoma Petroleum Industry*, p. 28; Clinton, "First Oil and Gas Well in Tulsa County," pp. 314-315; Franks, Lambert, and Tyson, *Early Oklahoma Oil*, p. 34.

[26]Franks, *The Oklahoma Petroleum Industry*, p. 28; Clinton, "First Oil and Gas Well in Tulsa County," p. 315; Franks, Lambert, and Tyson, *Early Oklahoma Oil*, p. 34.

[27]Franks, *The Oklahoma Petroleum Industry*, p. 28; Clinton, "First Oil and Gas Well in Tulsa County," p. 315.

[28]Franks, *The Oklahoma Petroleum Industry*, p. 28; Clinton, "First Oil and Gas Well in Tulsa County," pp. 315-316.

[29]Franks, *The Oklahoma Petroleum Industry*, p. 28; Clinton, "First Oil and Gas Well in Tulsa County," p. 323.

CHAPTER THREE
The Red Fork Oil Field

[1]Clinton, "First Oil and Gas Well in Tulsa County," p. 316; Franks, *The Oklahoma Petroleum Industry*, p. 30.

[2]Clinton, "First Oil and Gas Well in Tulsa County," pp. 313n, 316, 317n.; Gregory, "How *Glenn Pool* Oil Turned to Gold for Tulsa," p. 9.

[3]Franks, *The Oklahoma Petroleum Industry*, p. 30; C. B. Glasscock, *Then Came Oil: The Story of The Last Frontier* (New York: The Bobbs-Merrill Company, 1938), pp. 135-136.

[4]Franks, *The Oklahoma Petroleum Industry*, p. 30; Rister, *Oil!*, p. 84; Glasscock, *Then Came Oil*, p. 135; Gregory, "How *Glenn Pool* Oil Turned to Gold for Tulsa," p. 9.

[5]Franks, *The Oklahoma Petroleum Industry*, p. 30; Rister, *Oil!*, p. 84; Glasscock, *Then Came Oil*, p. 135; Gregory, "How Glenn Pool Oil Turned to Gold for Tulsa," p. 9.

[6]Rister, *Oil!*, pp. 84-85; Clinton, "First Oil and Gas Well in Tulsa County," pp. 318; Colcord, *The Autobiography of Charles Francis Colcord*, pp. 1; Gregory, "How Glenn Pool Oil Turned to Gold for Tulsa," pp. 9-10.

[7]Rister, *Oil!*, p. 84; Gregory, "How Glenn Pool Oil Turned to Gold for Tulsa," p. 9.

[8]Charles Francis Colcord, *The Autobiography of Charles Francis Colcord, 1859-1934* (Privately printed, 1970), pp. 202-203; Rister, *Oil!*, p. 84; Franks, *The Oklahoma Petroleum Industry*, p. 30; Glasscock, *Then Came Oil*, p. 135; Gregory, "How Glenn Pool Oil Turned to Gold for Tulsa," p. 9.

[9]Franks, *The Oklahoma Petroleum Industry*, p. 30; Rister, *Oil!*, p 84.

[10]Glasscock, *Then Came Oil*, p. 136; Rister, *Oil!*, pp. 84-86; Clinton, "First Oil and Gas Well in Tulsa County," pp. 318; Franks, *The Oklahoma Petroleum Industry*, p. 30; Gregory, "How Glenn Pool Oil Turned to Gold for Tulsa," p. 10.

[11]Franks, *The Oklahoma Petroleum Industry*, p. 31; Rister, *Oil!*, p. 85.

[12]Rister, *Oil!*, p. 85; Franks, *The Oklahoma Petroleum Industry*, p. 31.

[13]Franks, *The Oklahoma Petroleum Industry*, p. 32.

[14]Rister, *Oil!*, p. 86; Gregory, "How Glenn Pool Oil Turned to Gold for Tulsa," p. 10.

[15]Franks, *The Oklahoma Petroleum Industry*, p. 31; Rister, *Oil!*, p. 86.

[16]Rister, *Oil!*, p. 84; Colcord, *The Autobiography of Charles Francis Colcord*, pp. 192, 196, 198; Gregory, "How Glenn Pool Oil Turned to Gold for Tulsa," p. 10.

[17]Rister, *Oil!*, p. 84; Colcord, *The Autobiography of Charles Francis Colcord*, pp. 192, 196, 198; Gregory, "How Glenn Pool Oil Turned to Gold for Tulsa," p. 10; Clinton, "First Oil and Gas Well in Tulsa County," p. 316; Glasscock, *Then Came Oil*, p. 139.

[18]Rister, *Oil!*, p. 84; Colcord, *The Autobiography of Charles Francis Colcord*, pp. 192, 196, 198; Gregory, "How Glenn Pool Oil Turned to Gold for Tulsa," p. 10; Clinton, "First Oil and

Gas Well in Tulsa County," p. 316; Glasscock, *Then Came Oil*, p. 139.

[19]Rister, *Oil!*, p. 84; Colcord, *The Autobiography of Charles Francis Colcord*, pp. 192, 196, 198; Gregory, "How Glenn Pool Oil Turned to Gold for Tulsa," p. 10; Clinton, "First Oil and *ell in Tulsa County," p. 316; Glasscock, Then Came Oil*, p. 139.

[20]Clinton, "First Oil and Gas Well in Tulsa County," p. 316; Colcord, *The Autobiography of Charles Francis Colcord*, p. 198; Glasscock, *Then Came Oil*, p. 139.

[21]Colcord, *The Autobiography of Charles Francis Colcord*, pp. 198, 207; Rister, *Oil!*, p. 87; Glasscock, *Then Came Oil*, p. 139.

[22]Franks, *The Oklahoma Petroleum Industry*, p. 32.

[23]*Ibid.*; Seth K. Corden and W. B. Richards, comps., *The Oklahoma Red Book* (2 vols., Oklahoma City: Democrat Printing Company, 1912), Vol. I, p. 562.

[24]Corden and Richards, comp., *The Oklahoma Red Book, Vol. I*, pp. 562, 570; Franks, *The Oklahoma Petroleum Industry*, p. 32.

[25]Franks, *The Oklahoma Petroleum Industry*, p. 32; Rister, *Oil!*, pp. 86-87

[26]Franks, *The Oklahoma Petroleum Industry*, Rister, *Oil!*, pp. 86-87; *Tulsa Democrat* (Tulsa), August 2, 1901.

[27]Franks, *The Oklahoma Petroleum Industry*, p. 33.

[28]Corden and Richards, comp., *The Oklahoma Red Book*, Vol. I, p. 561; Franks, *The Oklahoma Petroleum Industry*, p. 33.

[29]Franks, *The Oklahoma Petroleum Industry*," p. 33; Rister, *Oil!*, pp. 86-87

[30]Franks, *The Oklahoma Petroleum Industry*, pp. 28; *Tulsa Democrat*, August 23, 1901.

[31]Franks, *The Oklahoma Petroleum Industry*, pp. 28; *Tulsa Democrat*, August 23, 1901; Litton, *History of Oklahoma*, Vol. II, p. 178; Rister, *Oil!*, p. 88.

[32]Clinton, "First Oil and Gas Well in Tulsa County," pp. 316-317.

[33]Dille, "Economic Geology: Oil and Gas in Tulsa County," p. 105; Clinton, "First Oil and Gas Well in Tulsa County," p. 317.

[34]Clinton, "First Oil and Gas Well in Tulsa County," pp. 323-324; Dille, "Economic Geology: Oil and Gas in Tulsa County," p. 105.

[35]Dille, "Economic Geology: Oil and Gas in Tulsa County," p. 122; Clinton, "The Beginning of the International Petroleum Exposition and Congress," pp. 479-480; Clinton, "First Oil and Gas Well in Tulsa County," pp. 317, 332; Gregory, "How Glenn Pool Oil Turned to Gold for Tulsa," p. 12.

[36]Dille, "Economic Geology: Oil and Gas in Tulsa County," p. 122; Clinton, "The Beginning of the International Petroleum Exposition and Congress," pp. 479-480; Clinton, "First Oil and Gas Well in Tulsa County," pp. 317, 332; Gregory, "How Glenn Pool Oil Turned to Gold for Tulsa," p. 12.

CHAPTER FOUR
Uncle Bill No. 1

[1]Morris, Goins, and McReynolds, *Historical Atlas of Oklahoma*, map 48; Corden and Richards, comps., *The Oklahoma Red Book*, Vol. I, pp. 426-436.

[2]Corden and Richards, comps., *The Oklahoma Red Book*, Vol. I, pp. 428-429; Franks, *The Oklahoma Petroleum Industry*, pp. 34-35.

[3]Franks, *The Oklahoma Petroleum Industry*, pp. 34-35; Franks, Lambert, Tyson, *Early Oklahoma Oil*, p. 36.

[4]Franks, *The Oklahoma Petroleum Industry*, pp. 34-35; Franks, Lambert, Tyson, *Early Oklahoma Oil*, p. 36.

[5]Franks, *The Oklahoma Petroleum Industry*, pp. 34-35; Franks, Lambert, Tyson, *Early Oklahoma Oil*, p. 36.

[6]Franks, *The Oklahoma Petroleum Industry*, pp. 34-35; Franks, Lambert, Tyson, *Early Oklahoma Oil*, p. 36.

[7]Franks, *Oklahoma Petroleum Industry*, p. 35; Paul B. Greig, *Geology of Pawnee County*, Oklahoma, *Oklahoma Geological Survey Bulletin No. 83* (November, 1959), pp. 6-7.

[8]Frank C. Greene, "Pawnee County," *Oil and Gas in Oklahoma, Oklahoma Geological Survey Bulletin No. 40* (July, 1930), p. 160.

[9]Litton, *History of Oklahoma*, Vol. I, p. 272; Morris, Goins, and McReynolds, Historical Atlas of Oklahoma, map 48.

[10]Litton, *History of Oklahoma*, Vol I, pp. 272-273, 399; Corden and Richards, comps., *Oklahoma Red Book*, Vol, I, pp. 464-465.

[11]Interview, Robert O. Jordan, September 26, 1983, Cleveland, Red Fork, and Glenn Pool Papers, Oklahoma Petroleum Collection, Archives, Oklahoma Heritage Association, Oklahoma City, Oklahoma.

[12]Corden and Richards, comps., *Oklahoma Red Book*, Vol, I, p. 490; Shirk, *Oklahoma Place Names*, p. 48; Litton, *History of Oklahoma*, Vol. I, pp. 399-401.

[13]Franks, *Oklahoma Petroleum Industry*, p. 40; "Map Cleveland, Boston, Lauderdale, and Maramec Oil and Gas Pools," *Petroleum and Natural Gas in Oklahoma, Oklahoma Geological Survey Bulletin No. 19*, Part II (April, 1917), Plate XXXV; *The Triangle* (Cleveland), May 27, 1904, June 3, 1904, and June 10, 1904; Rister, *Oil!*, p. 88; Litton, *History of Oklahoma*, Vol. II, p. 149.

[14]Franks, *The Oklahoma Petroleum Industry*, p. 40; *The Triangle*, June 3, 1904 and July 10, 1904.

[15]Franks, *The Oklahoma Petroleum Industry*, p. 40; *The Triangle*, June 3, 1904 and July 1, 1904.

[16]Franks, *The Oklahoma Petroleum Industry*, p. 50; *The Triangle*, June 3, 1904 and July 1, 1904.

[17]Franks, *The Oklahoma Petroleum Industry*, p. 50; *The Triangle*, June 3, 1904 and July 1, 1904.

[18]Franks, *The Oklahoma Petroleum Industry*, p. 50; *The Triangle*, July 8, 1904.

[19]Franks, *The Oklahoma Petroleum Industry*, p. 40; Rister, *Oil!*, p. 88; *The Triangle*, July 22, 1904 and July 29, 1904; *Petroleum and Natural Gas in Oklahoma, Oklahoma Geological Survey Bulletin No. 19*, Part II, p. 411; Greene, "Pawnee County," p. 177.

[20]Paul F. Lambert and Kenny A. Franks, "Oklahoma Oil Boom Towns," Oklahoma Petroleum Collection, Archives, Oklahoma Heritage Association, Oklahoma City, Oklahoma; Harold F. Williamson, Ralph L. Andreano, Arnold R. Daum, and Gilbert Klose, *The American Petroleum Industry* (2 vols., Evanston, Illinois: Northwestern University Press, 1959-1963), Vol. I, pp. 149-156.

[21]Lambert and Franks, "Oklahoma Oil Boom Towns"; Williamson, Andreano, Daum, and Klose, *The American Petroleum Industry*, Vol. I, pp. 149-156; Boone, *The Petroleum Dictionary*, p. 270; Lambert and Franks, "Oklahoma Oil Boom Towns"; Williamson, Andreano, Daum, and Klose, *The American Petroleum Industry*, Vol. I, pp. 149-156.

[22]Lambert and Franks, "Oklahoma Oil Boom Towns"; Williamson, Andreano, Daum, and Klose, *The American Petroleum Industry*, Vol. I, pp. 149-156; Rister, *Oil!*, pp. 88-89;

Litton, *History of Oklahoma*, Vol. II, p. 179; *The Triangle*, July 22, 1904 and July 29, 1904; Franks, Lambert, Tyson, *Early Oklahoma Oil*, p. 36.

[23]Franks, *The Oklahoma Petroleum Industry*, p. 40; Rister *Oil!*, pp. 88-89; Litton, *History of Oklahoma*, Vol. II, p. 179; *The Triangle*, July 22, 1904 and July 29, 1904; Franks, Lambert, Tyson, *Early Oklahoma Oil*, p. 36.

[24]Franks, Lambert, Tyson, *Early Oklahoma Oil*, p. 36; Franks, *The Oklahoma Petroleum Industry*, p. 40; Rister, *Oil!*, p. 88; *Petroleum and Natural Gas in Oklahoma, Oklahoma Geological Survey Bulletin No. 19*, p. 411.

[25]Franks, Lambert, Tyson, *Early Oklahoma Oil*, p. 37; Franks, *The Oklahoma Petroleum Industry*, p. 40; Rister, *Oil!*, p. 89.

[26]Franks, Lambert, Tyson, *Early Oklahoma Oil*, p. 37; Franks, *The Oklahoma Petroleum Industry*, p. 40; Rister, *Oil!*, p. 89; Interview, Jordan, September 26, 1983.

[27]Interview, Jordan, September 26, 1983.

[28]*Ibid.*

[29]*Ibid.*

[30]*Ibid.*; Interview, Robert L. Lumsford, III, September 26, 1983, Cleveland, Red Fork, and Glenn Pool Papers, Oklahoma Petroleum Collection, Archives, Oklahoma Heritage Association, Oklahoma City, Oklahoma.

[31]Interview, Jordan, September 26, 1983.

[32]*Ibid.*

[33]*Ibid.*

[34]*Ibid.*

[35]*Ibid.*

[36]Franks, Lambert, Tyson, *Early Oklahoma Oil*, p. 36; Franks, *The Oklahoma Petroleum Industry*, p. 40; Rister, *Oil!*, p. 89.

[37]Interview, Jordan, September 26, 1983.

[38]Franks, Lambert, Tyson, *Early Oklahoma Oil*, p. 36; Franks, *The Oklahoma Petroleum Industry*, p. 40; Rister, *Oil!*, p. 89.

[39]Mills-Bullard, comp., "Digest of Oklahoma Oil and Gas Fields," pp. 131; Greene, "Pawnee County," p. 177; *Petroleum and Natural Gas in Oklahoma, Oklahoma Geological Survey Bulletin No. 19*, pp. 411-412.

[40]Greene, "Pawnee County," p. 177; *Petroleum and Natural Gas in Oklahoma, Oklahoma Geological Survey Bulletin No. 19*, pp. 411-412.

[41]Franks, Lambert, Tyson, *Early Oklahoma Oil*, p. 36; Franks, *The Oklahoma Petroleum Industry*, p. 40; Rister, *Oil!*, p. 88; *Petroleum and Natural Gas in Oklahoma, Oklahoma Geological Survey Bulletin No. 19*, p. 411; Greene, "Pawnee County," p. 180.

[42]Greene, "Pawnee County," p. 177; *Petroleum and Natural Gas in Oklahoma, Oklahoma Geological Survey Bulletin No. 19*, pp. 411-412; Litton, *History of Oklahoma*, Vol. II, p. 179; Franks, Lambert, and Tyson, *Early Okahoma Oil*, p. 37; J. Stanley Clark, *The Oil Century: From the Drake Well to the Conservation Era* (Norman: University of Oklahoma Press, 1958), pp. 149-150.

[43]Greene, "Pawnee County," p. 177; *Petroleum and Natural Gas in Oklahoma, Oklahoma Geological Survey Bulletin No. 19*, pp. 411-412; Litton, *History of Oklahoma*, Vol. II, p. 179; Franks, Lambert, and Tyson, *Early Okahoma Oil*, p. 37; Clark, *The Oil Century*, pp. 149-150.

[44]Franks, *The Oklahoma Petroleum Industry*, pp. 73-74; Clark, *The Oil Century*, pp. 149-150.

[45]Rister, *Oil!*, p. 89; Franks, *The Oklahoma Petroleum Industry*, p. 40; Litton, *History of Oklahoma*, Vol. II, p. 179; Franks, Lambert, Tyson, *Early Oklahoma Oil*, p. 37.

[46]Franks, Lambert, Tyson, *Early Oklahoma Oil*, p. 37; Franks, *The Oklahoma Petroleum Industry*, pp. 40-41; Greene, "Pawnee County," pp. 177-185.

CHAPTER FIVE
Cleveland's Offspring

[1]Mills-Bullard, "Digest of Oklahoma Oil and Gas Fields," pp. 117-118.

[2]Interview, Jordan, September 26, 1983.

[3]*Ibid.*; Shirk, *Oklahoma Place Names*, p. 158; "Map Cleveland, Boston, Lauderdale, and Maramec Oil and Gas Pools".

[4]Greig, *Geology of Pawnee County*, pp. 143-145.

[5]*Ibid.*; Greene, "Pawnee County," p. 180; Mills-Bullard, "Digest of Oklahoma Oil and Gas Fields," pp. 227-228; *Petroleum and Natural Gas in Oklahoma, Oklahoma Geological Survey Bulletin No. 19*, pp. 412-413; "Pawnee County Map," Oklahoma Department of Transportation, Oklahoma City, Oklahoma.

[6]Greig, *Geology of Pawnee County*, pp. 143-145; Greene, "Pawnee County," p. 180; Mills-Bullard, "Digest of Oklahoma Oil and Gas Fields," p. 253; *Petroleum and Natural Gas in Oklahoma, Oklahoma Geological Survey Bulletin No. 19*, pp. 412-413; "Map Cleveland, Boston, Lauderdale, and Maramec Oil and Gas Pools".

[7]Greig, *Geology of Pawnee County*, pp. 143-145; Mills-Bullard, "Digest of Oklahoma Oil and Gas Fields," p. 226.

[8]Greene, "Pawnee County," pp. 181-182; Greig, *Geology of Pawnee County*, pp. 143-145.

[9]Everett Carpenter, "The Morrison Field, Pawnee County Oklahoma," *Oil and Gas in Oklahoma, Oklahoma Geological Survey Bulletin No. 40* (July, 1930), pp. 184-195.

[10]Greig, *Geology of Pawnee County*, pp. 143-145; "Pawnee County Map"; *Petroleum and Natural Gas in Oklahoma, Oklahoma Geological Survey Bulletin No. 19*, p. 412.

[11]*Petroleum and Natural Gas in Oklahoma, Oklahoma Geological Survey Bulletin No. 19*, p. 413; Greig, *Geology of Pawnee County*, pp. 143-145; Greene, "Pawnee County," pp. 178-179, 181; Mills-Bullard, "Digest of Oklahoma Oil and Gas Fields," p. 183.

[12]*Petroleum and Natural Gas in Oklahoma, Oklahoma Geological Survey Bulletin No. 19*, p. 412; Greig, *Geology of Pawnee County*, pp. 143-145; Greene, "Pawnee County," pp. 178, 181; Mills-Bullard, "Digest of Oklahoma Oil and Gas Fields," pp. 176-177.

[13]Greig, *Geology of Pawnee County*, pp. 143-145; Greene, "Pawnee County," p. 178; Mills-Bullard, "Digest of Oklahoma Oil and Gas Fields," p. 164.

[14]Greig, *Geology of Pawnee County*, pp. 143-145; Greene, "Pawnee County," p. 181.

[15]Greene, "Pawnee County," p. 178; Mills-Bullard, "Digest of Oklahoma Oil and Gas Fields," pp. 179-180; Greig, *Geology of Pawnee County*, pp. 143-145.

[16]Greig, *Geology of Pawnee County*, pp. 143-145; *Petroleum and Natural Gas in Oklahoma, Oklahoma Geological Survey Bulletin No. 19*, pp. 412-413.

[17]Greig, *Geology of Pawnee County*, pp. 143-145.

[18]*Ibid.*

[19]*Ibid.*

[20]*Ibid.*; Greene, "Pawnee County," pp. 179, 181; Mills-Bullard, "Digest of Oklahoma Oil and Gas Fields," pp. 147-148.

[21]Greene, "Pawnee County," pp. 177, 179-181.

[22]*Ibid.*

[23]J. Paul Getty, *As I See It: The Autobiography of J. Paul Getty*

(Englewood Cliffs, New Jersey: Prentice-Hall, Inc., 1976), pp. 20-21; Max Gunther, *The Very, Very Rich and How They Got That Way* (Chicago: Playboy Press, 1972), pp. 107-108; Ralph Hewins, *The Richest American: J. Paul Getty* (New York: E. P. Dutton & Co., Inc., 1960), pp. 41-42.

[24]Getty, *As I See It*: pp. 20-22; Hewins, *The Richest American*, pp. 41-42.

[25]Getty, *As I See It*: pp. 20-22; Hewins, *The Richest American*, pp. 42-43.

[26]Getty, *As I See It*, p. 22.

[27]*Ibid.*, pp. 22-24; Hewins, *The Richest American*, pp. 46-48.

[28]Getty, *As I See It*, pp. 22-24; Hewins, *The Richest American*, pp. 46-47.

[29]Getty, *As I See It*, pp. 24-25.

[30]J. Paul Getty, *My Life and My Fortunes* (New York: Duell, Sloan and Pearce, 1963), p. 35.

[31]Getty, *As I See It*, pp. 25-26.

[32]*Ibid.*, p. 26-27; Getty, *My Life and My Fortunes*, pp. 33-36.

[33]Getty, *My Life and My Fortunes*, pp. 36-37; Hewins, *The Richest American*, pp. 50-51; J. Paul Getty, *The History of the Oil Business of George F. and J. Paul Getty from 1903 to 1939* (Privately printed, 1941), p. 37.

[34]Getty, *My Life and My Fortunes*, pp. 36-37; Hewins, *The Richest American*, pp. 50-51; Getty, *The History of the Oil Business of George F. and J. Paul Getty*, p. 37.

[35]Getty, *My Life and My Fortunes*, pp. 36-37; Hewins, *The Richest American*, pp. 50-51; Getty, *The History of the Oil Business of George F. and J. Paul Getty*, p. 37.

[36]Getty, *My Life and My Fortunes*, pp. 36-37; Hewins, *The Richest American*, pp. 50-51; Getty, *The History of the Oil Business of George F. and J. Paul Getty*, p. 37.

[37]Hewins, *The Richest American*, pp. 50-51; Getty, *The History of the Oil Business of George F. and J. Paul Getty*, pp. 37-38.

[38]Getty, *As I See It*, pp. 26-27.

[39]*Ibid.*

[40]*Ibid.*, pp. 27-30, 34; Gunther, *The Very, Very Rich And How They Got That Way*, p. 112-113; Hewins, *The Richest American*, p. 65; Getty, *My Life and My Fortunes*, p. 55.

[41]Getty, *As I See It*, pp. 27-30, 34; Gunther, *The Very, Very Rich And How They Got That Way*, p. 112-113; Hewins, *The Richest American*, p. 65; Getty, *My Life and My Fortunes*, p. 55.

[42]Getty, *As I See It*, pp. 34, 40, 352-353. Among the Getty companies are: the Associated Oil Company (California); Eastern Gasoline Retailers, Inc. (Delaware); Getty Iran, Ltd. (Delaware); Getty Minerals Company, Limited (Canada); Getty Mines, Limited (Canada); Getty Mining Northeast Limited (Canada); Getty Mining Northwest, Limited (Canada); Getty Mining Pacific, Limited (Canada); Getty Mining Proprietary Limited (Australia); Getty Oil (Angola) Limited (Delaware); Getty Oil (Artic) Ltd. (Canada); Getty Oil (Baffin Island) Ltd. (Canada); Getty Oil (Canadian Operations) Ltd. (Canada); Getty Oil (Congo) Inc. (Delaware); Getty Oil (Eagle Basin) Ltd. (Alberta); Getty Oil (Foothills) Ltd. (Canada); Getty Oil (Guatemala) Inc. (Delware); Getty Oil (Ivory Coast) Inc. (Delaware); Getty Oil (Mackenzie Delta) Ltd. (Canada); Getty Oil (Maritime) Ltd. (Canada); Getty Oil (Mossel Bay) Ltd. (Delaware); Getty Oil (Netherlands) Ltd. (Delaware); Getty Oil (North Star) Ltd. (Canada); Getty Oil (Northwest Territories) Ltd. (Canada); Getty Oil (Norway) A/S (Delaware); Getty Oil (N.S.) Ltd. (Canada); Getty Oil (Pacific) Ltd. (Canada); Getty Oil (Peru) Inc. (Delaware); Getty Oil (Philippines) Inc. (Philippines); Getty Oil (Plains) Ltd. (Canada); Getty Oil (Plateau) Ltd. (Canada); Getty Oil (Sumatra) Inc. (Delaware); Getty Oil (Trinidad and Tobago) Inc. (Delaware); Getty Oil (United Kingdom) Ltd. (Delaware); Getty Oil (Walvis Bay) Ltd. (Delaware); Getty Oil Company (Eastern operations) Inc. (Delaware); Getty Oil Company, Ltd. (Alberta); Getty Oil Company of Spain, S.A. (Delaware); Getty Oil International (Argentina) Inc. (Liberia); Getty Oil International (England) Limited (United Kingdom); Getty Oil International (Hebrides) Limited (United Kingdom); Getty Oil Intenational (Indonesia) Inc. (Liberia); Getty Oil International (Irish Sea) Limited (United Kingdom); Getty Oil Intrnational (Scotland) Limited (United Kingdom); Getty Oil International (Spain) S.A. (Delaware); Getty Petroleum Company (Liberia); Getty Pipe Company (Pennsylvania); Getty Pipe Line Company (Texas); Hemisphere Transportation Corporation (Liberia); Industrial Royalties Company (California); Liberian Operations Inc. (Liberia); Liberian Operations Ltd. (England); Minnehoma Corporation (Delaware); Minnehoma Land and Farming Company (California); Pacific Western Oil Corporation (Delaware); and Sutton Place Property Company, Limited (England) as wholly owned subsidiaries. Also Getty Oil Company owns 40 percent of Mitsubishi Oil Company, Limited (Japan) and controls as consolidated companies Mission Corporation (Nevada); Skelly Oil Company (Delaware); and Nuclear Fuel Service, Inc. (Maryland). On January 6, 1984, Texaco announced that it had acquired Getty Oil.

[43]Franks, *The Oklahoma Petroleum Industry*, pp. 185-186; Greig, *Geology of Pawnee County*, pp. 143-145.

[44]Greig, *Geology of Pawnee County*, pp. 143-145.

[45]*Ibid.*

[46]*Ibid.*, p. 145.

[47]*Ibid.*, pp. 144-145

[48]Oklahoma Corporation Commission, Oil & Gas Conservation Division, "Annual Summary by Quarter of Liquids & Gas Runs by Counties, 1980," Oklahoma Publications Clearinghouse, Oklahoma Department of Libraries, Oklahoma City, Oklahoma; Oklahoma Corporation Commission, Oil & Gas Conservation Division, "January, 1983, Liquids and Gas Runs by Counties," Oklahoma Publications Clearinghouse, Oklahoma Department of Libraries, Oklahoma City, Oklahoma.

[49]Franks, *The Oklahoma Petroleum Industry*, p. 35.

CHAPTER SIX
Ida Glenn No. 1

[1]Franks, *The Oklahoma Petroleum Industry*, p. 41; Franks, Lambert, Tyson, *Early Oklahoma Oil*, p. 37; Gregory, "How Glenn Pool Oil Turned to Gold for Tulsa," pp. 16-20; J. Vere Frazier, Jr., "History of the Glennpool Oil Field" (M.A. Thesis, University of Oklahoma, 1951).

[2]Gregory, "How Glenn Pool Oil Turned to Gold for Tulsa," p. 20; Frazier, Jr., "History of the Glennpool Oil Field, pp. 8-9.

[3]Gregory, "How Glenn Pool Oil Turned to Gold for Tulsa," pp. 20-22; Franks, *The Oklahoma Petroleum Industry*, p. 41; Franks, Lambert, Tyson, *Early Oklahoma Oil*, p. 37; Frazier, Jr., "History of the Glennpool Oil Field, pp. 8-9.

[4]Gregory, "How Glenn Pool Oil Turned to Gold for Tulsa,"

pp. 20-22, 25; Franks, *The Oklahoma Petroleum Industry*, p. 41; Franks, Lambert, Tyson, *Early Oklahoma Oil*, p. 37; Frazier, Jr., "History of the Glennpool Oil Field, p. 12.

[5]Gregory, "How Glenn Pool Oil Turned to Gold for Tulsa," pp. 22.

[6]Gregory, "How Glenn Pool Oil Turned to Gold for Tulsa," p. 20; Franks, *The Oklahoma Petroleum Industry*, p. 41; Franks, Lambert, Tyson, *Early Oklahoma Oil*, p. 37.

[7]Gregory, "How Glenn Pool Oil Turned to Gold for Tulsa," pp. 20-21.

[8]Frank Galbreath, *Glenn Pool...and a Little Oil Town of Yesteryear* (Privately printed, 1978), p, 93.

[9]*Ibid.*

[10]Gregory, "How Glenn Pool Oil Turned to Gold for Tulsa," pp. 21-22; Franks, Lambert, Tyson, *Early Oklahoma Oil*, p. 37; Frazier, Jr., "History of the Glennpool Oil Field, p. 11; Interview, James W. Hubbard, September 27-28, 1983, Cleveland, Red Fork, and Glenn Pool Papers, Oklahoma Petroleum Collection, Archives, Oklahoma Heritage Association.

[11]Shirk, *Oklahoma Place Names*, p. 117; Gregory, "How Glenn Pool Oil Turned to Gold for Tulsa," pp. 22.

[12]Franks, *The Oklahoma Petroleum Industry*, pp. 41-42; Gregory, "How Glenn Pool Oil Turned to Gold for Tulsa," p. 22; Franks, Lambert, Tyson, *Early Oklahoma Oil*, pp. 37-38; Frazier, Jr., "History of the Glennpool Oil Field, p. 12.

[13]Gregory, "How Glenn Pool Oil Turned to Gold for Tulsa," p. 22; Franks, *The Oklahoma Petroleum Industry*, p. 41; Galbreath, *Glenn Pool*, p. 97.

[14]Franks, Lambert, Tyson, *Early Oklahoma Oil*, pp. 37-38.

[15]*Ibid.*

[16]*Ibid.*, pp. 37-38; Franks, *The Oklahoma Petroleum Industry*, p. 41; Gregory, "How Glenn Pool Oil Turned to Gold for Tulsa," p. 22; Galbreath, *Glenn Pool*, p. 98.

[17]Gregory, "How Glenn Pool Oil Turned to Gold for Tulsa," p. 22; Franks, Lambert, Tyson, *Early Oklahoma Oil*, pp. 37-38; Frazier, Jr., "History of the Glennpool Oil Field, p. 13; Galbreath, *Glenn Pool*, p. 98.

[18]Gregory, "How Glenn Pool Oil Turned to Gold for Tulsa," p. 22; Franks, Lambert, Tyson, *Early Oklahoma Oil*, pp. 37-38; "Tulsa County Map"; "Creek County Map No. 2," Oklahoma Department of Transportation, Oklahoma City, Oklahoma.

[19]Franks, Lambert, Tyson, *Early Oklahoma Oil*, p. 38; Gregory, "How Glenn Pool Oil Turned to Gold for Tulsa," pp. 22-25; Franks, *The Oklahoma Petroleum Industry*, pp. 41-42.

[20]Franks, *The Oklahoma Petroleum Industry*, p. 42; Gregory, "How Glenn Pool Oil Turned to Gold for Tulsa," p. 27; Franks, Lambert, Tyson, *Early Oklahoma Oil*, pp. 37-38; Frazier, Jr., "History of the Glennpool Oil Field, p. 13.

[21]Frazier, Jr., "History of the Glennpool Oil Field," p. 27.

[22]Gregory, "How Glenn Pool Oil Turned to Gold for Tulsa," pp. 25-26.

[23]Franks, Lambert, Tyson, *Early Oklahoma Oil*, p. 38; Franks, Lambert, Tyson, *Early Oklahoma Oil*, p. 38; Frazier, Jr., "History of the Glennpool Oil Field," pp. 26-27; Gregory, "How Glenn Pool Oil Turned to Gold for Tulsa," p. 30.

[24]Franks, Lambert, Tyson, *Early Oklahoma Oil*, p. 38; Frazier, Jr., "History of the Glennpool Oil Field," p. 20; Franks, *The Oklahoma Petroleum Industry*, p. 43; Galbreath, *Glenn Pool*, p. 94.

[25]Franks, Lambert, Tyson, *Early Oklahoma Oil*, p. 38; Frazier, Jr., "History of the Glennpool Oil Field," p. 20; Franks, *The Oklahoma Petroleum Industry*, p. 43; Galbreath, *Glenn Pool*, p. 94.

[26]Forbes, "Orgin and Early Development of the Oil Industry in Oklahoma," pp. 95-96.

[27]Shirk, *Oklahoma Place Names*, p. 112; Frazier, "History of Glenn Pool," p. 64.

[28]Franks, *The Oklahoma Petroleum Industry*, p. 43; Gregory, "How Glenn Pool Oil Turned to Gold for Tulsa," p. 30; Franks, Lambert, Tyson, *Early Oklahoma Oil*, p. 38; Galbreath, Glenn Pool, pp. 113-114; Frazier, Jr., "History of the Glennpool Oil Field," pp. 20-21.

[29]Frazier, Jr., "History of the Glennpool Oil Field," pp. 20-21; Gregory, "How Glenn Pool Oil Turned to Gold for Tulsa," p. 30.

[30]Galbreath, *Glenn Pool*, pp. 113-114; Frazier, Jr., "History of the Glennpool Oil Field," pp. 20-21; Gregory, "How Glenn Pool Oil Turned to Gold for Tulsa," p. 30.

[31]Frazier, Jr., "History of the Glennpool Oil Field," pp. 20-21; Gregory, "How Glenn Pool Oil Turned to Gold for Tulsa," p. 30.

[32]Frazier, Jr., "History of the Glennpool Oil Field," pp. 21-22.

[33]Franks, *The Oklahoma Petroleum Industry*, p. 43; Gregory, "How Glenn Pool Oil Turned to Gold for Tulsa," p. 30; Franks, Lambert, Tyson, *Early Oklahoma Oil*, p. 38.

[34]Franks, Lambert, Tyson, *Early Oklahoma Oil*, p. 38.

[35]*Ibid.*; Gregory, "How Glenn Pool Oil Turned to Gold for Tulsa," pp. 26-27, 32.

[36]Gregory, "How Glenn Pool Oil Turned to Gold for Tulsa," pp. 26-27, 32.

[37]*Ibid.*, p. 29; Frazier, Jr., "History of the Glennpool Oil Field," pp. 14-17, 20; Franks, Lambert, Tyson, *Early Oklahoma Oil*, p. 38.

[38]Frazier, Jr., "History of the Glennpool Oil Field," pp. 14-17; Gregory, "How Glenn Pool Oil Turned to Gold for Tulsa," p. 29.

[39]Franks, Lambert, Tyson, *Early Oklahoma Oil*, p. 38; Frazier, Jr., "History of the Glennpool Oil Field," pp. 16-17, 23; Franks, *The Oklahoma Petroleum Industry*, p. 42.

[40]Franks, Lambert, Tyson, *Early Oklahoma Oil*, p. 38; Frazier, Jr., "History of the Glennpool Oil Field," pp. 16-17, 23; Franks, *The Oklahoma Petroleum Industry*, p. 42.

[41]Frazier, Jr., "History of the Glennpool Oil Field," p. 15; Franks, *The Oklahoma Petroleum Industry*, p. 42; Rister, *Oil!*, p. 91.

[42]Frazier, Jr., "History of the Glennpool Oil Field," p. 15; Franks, *The Oklahoma Petroleum Industry*, p. 42.

[43]Frazier, Jr., "History of the Glennpool Oil Field," p. 20; Gregory, "How Glenn Pool Oil Turned to Gold for Tulsa," p. 29.

[44]American Petroleum Institute, *Petroleum Facts and Figures, 1950* (New York: American Petroleum Institute, 1951), pp. 145-149; Franks, *The Oklahoma Petroleum Industry*, p. 46.

CHAPTER SEVEN
Flush Production

[1]Franks, Lambert, Tyson, *Early Oklahoma Oil*, p. 39.

[2]Franks, *The Oklahoma Petroleum Industry*, p. 44; Sinclair Oil Corporation, *A Great Name in Oil: Sinclair Through Fifty Years* (New York: McGraw-Hill, Inc., 1966), pp. 14-15.

[3]Franks, *The Oklahoma Petroleum Industry*, p. 44; Sinclair Oil Corporation, *A Great Name in Oil*, pp. 14-15.

[4]Franks, *The Oklahoma Petroleum Industry*, p. 44; Sinclair Oil Corporation, *A Great Name in Oil*, pp. 14-18.

[5]Rister, *Oil!*, p. 93, 93n; Getty, *As I See It*, pp. 45-46. In 1916, Tidewater changed it name to the Tidal Oil Company, and then in 1932 the Tide Water Oil Company of Delaware. Four years later, in 1936 it was merged into the Tide Water Associated Oil Company.

[6]Rister, *Oil!*, p. 93, 93n; Gregory, "How Glenn Pool Oil Turned to Gold for Tulsa," p. 39.

[7]Franks, *The Oklahoma Petroleum Industry*, p. 44.

[8]W. L. Connelly, *The Oil Business As I Saw It: Half a Century with Sinclair* (Norman: University of Oklahoma Press, 1954), p. 29.

[9]Connelly, *The Oil Business As I Saw It*, p. 29; ; Sapulpa Historical Society, *Sapulpa, Ok 74066* (2 vols., Sapulpa: Sapulpa Historical Society, 1979-1981), Vol. I, pp. 103-105; Boone, *The Petroleum Dictionary*, pp. 261, 321.

[10]Sapulpa Historical Society, *Sapulpa, Ok 74066*, Vol. I, p. 105.

[11]Connelly, *The Oil Business As I Saw It*, p. 29.

[12]Clark, *The Oil Century*, pp. 149-150.

[13]Boone, *The Petroleum Dictionary*, pp. 185-186.

[14]Galbreath, *Glenn Pool*, pp. v, 108.

[15]Frazier, "History of the Glennpool Oil Field," pp. 18-19; Mills-Bullard, comp., "Digest of Oklahoma Oil and Gas Fields," pp. 160-161.

[16]Franks, Lambert, Tyson, *Early Oklahoma Oil*, p. 39; Rister, *Oil!*, p. 90; Frazier, "History of the Glennpool Oil Field," pp. 22, 18; Gregory, "How Glenn Pool Oil Turned to Gold for Tulsa," p. 39.

[17]Rister, *Oil!*, p. 91; Franks, Lambert, Tyson, *Early Oklahoma Oil*, p. 39; Frazier, "History of the Glennpool Oil Field," pp. 24-25, 28-29, 37.

[18]Rister, *Oil!*, p. 91; Franks, Lambert, Tyson, *Early Oklahoma Oil*, p. 39; Frazier, "History of the Glennpool Oil Field," pp. 24-25, 28-29, 37.

[19]Rister, *Oil!*, p. 90; Franks, Lambert, Tyson, *Early Oklahoma Oil*, p. 39.

[20]Franks, *The Oklahoma Petroleum Industry*, p. 46; Gregory, "How Glenn Pool Oil Turned to Gold for Tulsa," p. 32.

[21]Franks, *The Oklahoma Petroleum Industry*, p. 46; Gregory, "How Glenn Pool Oil Turned to Gold for Tulsa," p. 32.

[22]Gregory, "How Glenn Pool Oil Turned to Gold for Tulsa," p. 32; Franks, Lambert, Tyson, *Early Oklahoma Oil*, p. 39; Rister, *Oil!*, p. 91.

[23]Rister, *Oil!*, p. 91; Frazier, "History of the Glenn Pool Oil Field," p. 18.

[24]Galbreath, *Glenn Pool*, pp. 103-106.

[25]*Ibid.*, p. 106.

[26]Rister, *Oil!*, p. 92; Marquis James, *The Texaco Story: The First Fifty Years, 1902-1952* (N.p., The Texas Company, 1953), p. 32; Franks, *The Oklahoma Petroleum Industry*, p. 45.

[27]Rister, *Oil!*, p. 92; James, *The Texaco Story*, pp. 32-33; ; Frazier, "History of the Glennpool Oil Field," pp. 39-40, 53.

[28]James, *The Texaco Story*, pp. 32-33; Frazier, "History of Glennpool Oil Field," pp. 39-40.

[29]James, *The Texaco Story*, pp. 32-33; Frazier, "History of Glennpool Oil Field," p. 39.

[30]Craig Thompson, *Gulf Since Spindletop: A Human Story of Gulf's First Half-Century* (Privately printed, n.d.), pp. 25-26; Rister, *Oil!*, p. 92.

[31]Thompson, *Gulf Since Spindletop*, p. 26.

[32]*Ibid.*; Frazier, "History of Glennpool Oil Field," p. 40; Frazier, "History of the Glennpool Oil Field," pp. 40-41.

[33]Thompson, *Gulf Since Spindletop*, p. 26; Frazier, "History of Glennpool Oil Field," p. 40; Frazier, "History of the Glennpool Oil Field," pp. 40-41.

[34]Thompson, *Gulf Since Spindletop*, p. 26; Frazier, "History of the Glennpool Oil Field," pp. 40-41.

[35]Thompson, *Gulf Since Spindletop*, pp. 26-27.

[36]Rister, *Oil!*, p. 92; Thompson, *Gulf Since Spindletop*, p. 96.

[37]Franks, *The Oklahoma Petroleum Industry*, p. 44; Sinclair Oil Corporation, *A Great Name in Oil*, pp. 17-39.

[38]Rister, *Oil!*, p. 92; Robert A. Ferrell, "The Oklahoma Natural Gas Company: A Study in Industrial Geography" (M.A. Thesis, Norman, 1964), p. 12.

[39]Ferrell, "The Oklahoma Natural Gas Company," pp. 13-14.

[40]Rister, *Oil!*, p. 92; Ferrell, "The Oklahoma Natural Gas Company," p. 14.

[41]Franks, *The Oklahoma Petroleum Industry*, p. 46; Franks, Lambert, Tyson, *Early Oklahoma Oil*, pp. 39-40.

[42]Franks, *The Oklahoma Petroleum Industry*, pp. 45-46; Franks, Lambert, Tyson, *Early Oklahoma Oil*, pp. 39-40; Frazier, "History of the Glennpool Oil Field," p. 56; Gregory, "How Glenn Pool Oil Turned to Gold for Tulsa," p. 39.

[43]Franks, *The Oklahoma Petroleum Industry*, p. 49; Franks, Lambert, Tyson, *Early Oklahoma Oil*, p. 40.

[44]Franks, Lambert, Tyson, *Early Oklahoma Oil*, p. 40; Corden and Richards, comps., *The Oklahoma Red Book*, Vol. I, p. 548; Franks, *The Oklahoma Petroleum Industry*, pp. 50-52.

[45]Franks, *The Oklahoma Petroleum Industry*, pp. 52-53; Franks, Lambert, Tyson, *Early Oklahoma Oil*, p. 40.

[46]Galbreath, *Glenn Pool*, p. 120; Franks, Lambert, Tyson, *Early Oklahoma Oil*, pp. 40-41.

[47]Franks, *The Oklahoma Petroleum Industry*, p. 54; Franks, Lambert, Tyson, *Early Oklahoma Oil*, p. 41.

[48]Franks, *The Oklahoma Petroleum Industry*, pp. 54; Franks, Lambert, Tyson, *Early Oklahoma Oil*, pp. 40-41.

[49]Franks, *The Oklahoma Petroleum Industry*, pp. 54-55; Franks, Lambert, Tyson, *Early Oklahoma Oil*, pp. 40-41

[50]Franks, *The Oklahoma Petroleum Industry*, pp. 45-46; Franks, Lambert, Tyson, *Early Oklahoma Oil*, p. 41; Sinclair Oil Corporation, *A Great Name in Oil*, p. 16; Gregory, "How Glenn Pool Oil Turned to Gold for Tulsa," p. 39; American Petroleum Institute, *Petroleum Facts and Figures, Ninth Edition, 1950*, pp. 145-149, 156; Litton, *History of Oklahoma*, p. 180.

CHAPTER EIGHT
Glenn Pool's Boom Towns

[1]Frazier, "History of Glenn Pool," p. 66; Shirk, *Oklahoma Place Names*, pp. 27, 89, 112, 117, 144, 184, 187; Rister, *Oil!*, p. 91; Sapulpa Historical Society, *Sapulpa, Ok 74066*, Vol. I, pp. 135-136.

[2]Frazier, "History of Glenn Pool," pp. 65-66; Shirk, *Oklahoma Place Names*, p. 89.

[3]Frazier, "History of Glenn Pool," pp. 65-66; Shirk, *Oklahoma Place Names*, p. 89.

[4]Frazier, "History of Glenn Pool," pp. 66-67; Walter G. McComas, "Sapulpa, The Metropolis of the World's Greatest Oil Museum," *Sturm's Oklahoma Magazine,* Vol. IV, No. 2 [3?] (May, 1908), p. 24.

[5]Frazier, "History of Glenn Pool," pp. 66-67; McComas, "Sapulpa, The Metropolis of the World's Greatest Oil Museum," p. 24; Shirk, *Place Names*, p. 187.

[6]Gregory, "How Glenn Pool Oil Turned to Gold for Tulsa," pp. 29-30; Sapulpa Historical Society, *Sapulpa, Ok 74066,* Vol. II, pp. 282-285; Charles G. Forbes, "The Origin and Early Development of the Oil Industry in Oklahoma" (Ph.D. Dissertation, University of Oklahoma, 1939), p. 84; Frazier, "History of Glenn Pool," pp. 66-69; McComas, "Sapulpa, The Metropolis of the World's Greatest Oil District," p. 28.

[7]Gregory, "How Glenn Pool Oil Turned to Gold for Tulsa," pp. 29-30; Sapulpa Historical Society. *Sapulpa, Ok 74066,* Vol. II, pp. 282-285; Charles G. Forbes, "The Origin and Early Development of the Oil Industry in Oklahoma," p. 84; Frazier, "History of Glenn Pool," pp. 66-69.

[8]Interview, Jack Dillon, "Indian-Pioneer History," Grant Foreman, ed., 113 vols., Indian Archives, Oklahoma Historical Society, Oklahoma City, Oklahoma, Vol. III, p. 41.

[9]Forbes, "The Origin and Early Development of the Oil Industry in Oklahoma," p. 84; Frazier, "History of Glenn Pool," pp. 66-69; Gregory, "How Glenn Pool Oil Turned to Gold for Tulsa," pp. 29-30; Sapulpa Historical Society, *Sapulpa, Ok 74066,* Vol. II, p. 285; McComas, "Sapulpa, The Metropolis of the World's Greatest Oil District," p. 28.

[10]Gregory, "How Glenn Pool Oil Turned to Gold for Tulsa," p. 29; McComas, "Sapulpa, The Metropolis of the World's Greatest Oil District," pp. 24-25, 28; Frazier, "History of Glenn Pool," pp. 66-69; Sapulpa Historical Society, *Sapulpa, Ok 74066,* Vol. II, p. 285.

[11]Sapulpa Historical Society, *Sapulpa, Ok 74066,* Vol. II, p. 162.

[12]Frazier, "History of Glenn Pool," pp. 66-69; Gregory, "How Glenn Pool Oil Turned to Gold for Tulsa," p. 29; Sapulpa Historical Society, *Sapulpa, Ok 74066,* Vol. II, pp. 27-28, 61; McComas, "Sapulpa, The Metropolis of the World's Greatest Oil District," p. 28.

[13]Sapulpa Historical Society, *Sapulpa, Ok 74066,* Vol. I, pp. 108-109.

[14]Frazier, "History of Glenn Pool," pp. 66-69; Gregory, "How Glenn Pool Oil Turned to Gold for Tulsa," p. 34; *Sunday Oklahoman* (Oklahoma City), July 17, 1983.

[15]Frazier, "History of Glenn Pool," pp. 66-69; Gregory, "How Glenn Pool Oil Turned to Gold for Tulsa," p. 34; *Sunday Oklahoman,* July 17, 1983.

[16]Sapulpa Historical Society, *Sapulpa Ok 74066,* Vol. II, pp. 48, 109, 288.

[17]Connelly, *The Oil Business As I Saw It,* p. 32.

[18]*Ibid.*

[19]*Ibid.*

[20]*Ibid.*

[21]*Ibid.*, pp. 32-33.

[22]*Ibid.*

[23]*Ibid.*

[24]*Ibid.*

[25]Sapulpa Historical Society, *Sapulpa Ok 74066,* Vol. II, p.

[26]*Ibid.*, pp. 48, 153.

[27]*Ibid.*, 48, 153.

[28]Franks and Lambert, "Oklahoma Oil Boom Towns"; Frazier, "History of Glenn Pool," pp. 58-62; Shirk, *Oklahoma Place Names,* p. 117; Rister, *Oil!,* p. 91; Interview, Dillon, p. 40.

[29]Franks and Lambert, "Oklahoma Oil Boom Towns"; Frazier, "History of Glenn Pool," pp. 58-62; Shirk, *Oklahoma Place Names,* p. 117; Rister, *Oil!,* p. 91; Interview, Dillon, p. 40.

[30]Sapulpa Historical Society, Sapulpa, Ok 74066, pp. 126-130.

[31]*Ibid.*

[32]*Ibid.*

[33]*Ibid.*

[34]*Ibid.*

[35]*Ibid.*

[36]*Ibid.*

[37]*Ibid.*

[38]*Ibid.*

[39]Frazier, "History of Glenn Pool," pp. 58-62, 88; Rister, *Oil!,* p. 91.

[40]Rister, *Oil!,* p. 91; Frazier, "History of Glenn Pool," pp. 59-60; Franks, *The Oklahoma Petroleum Industry*, p. 202.

[41]David Craighead, "Glenn Pool: A Forest of Derricks," Oklahoma Today, Vol. XVI, No. 1 (Winter 1965-1966), p. 11; Frazier, "History of Glenn Pool," pp. 59-60.

[42]Frazier, "History of Glenn Pool," pp. 25, 59-60.

[43]*Ibid.*, p. 61.

[44]Frazier, "History of Glenn Pool," pp. 61; Rister, *Oil!,* p. 91.

[45]Frazier, "History of Glenn Pool, p. 62; Rister, *Oil!,* p. 91.

[46]Gregory, "How Glenn Pool Oil Turned to Gold for Tulsa," p. 39; Franks, *The Oklahoma Petroleum Industry*, p. 95; Frazier, "History of Glenn Pool, pp. 62-63; Gerald Forbes, "Southwestern Oil Boom Towns," *The Chronicles of Oklahoma,* Vol. XVII, No. 4 (December, 1939), pp. 396-397.

[47]Gregory, "How Glenn Pool Oil Turned to Gold for Tulsa," p. 39; Franks, *The Oklahoma Petroleum Industry*, p. 95; Frazier, "History of Glenn Pool, pp. 62-63; Interview, Dillon, p. 37.

[48]Interview, Dillion, pp. 37-38.

[49]Sapulpa Historical Society, *Sapulpa, Ok 74066*, p. 131.

[50]*Ibid.*

[51]*Ibid.*

[52]*Ibid.*

[53]Franks and Lambert, "Oklahoma Oil Boom Towns".

[54]Interview, Dillion, p. 38; Sapulpa Historical Society, *Sapulpa, Ok 74066,* Vol. I, p. 78.

[55]Interview, Dillon, p. 38; Sapulpa Historical Society, *Sapulpa, Ok 74066*, pp. 84-85.

[56]Franks and Lambert, "Oklahoma Oil Boom Towns"; Frazier, "History of Glenn Pool," pp. 62-63; Forbes, "Southwestern Oil Boom Towns," p. 398.

[57]Franks and Lambert, "Oklahoma Oil Boom Towns"; Frazier, "History of Glenn Pool," p. 63; Department of Economic and Community Affairs, Oklahoma State Data Center, "1980 Census Data: Population and Household Characteristics," United States Documents Regional Depository, Oklahoma Department of Libraries, Oklahoma City, Oklahoma.

[58]Shirk, *Oklahoma Place Names*, pp. 27, 184; Sapulpa Historical Society, *Sapulpa, Ok 74066*, pp. 135-136.

[59]Shirk, *Oklahoma Place Names*, pp. 27, 184; Interview, Hubbard, September 26-27, 1983.

[60]Shirk, *Oklahoma Place Names*, p. 144; Frazier, "History of Glenn Pool," p. 63.

[61]Frazier, "History of Glenn Pool," pp. 63-64; Gregory, "How Glenn Pool Oil Turned to Gold for Tulsa," p. 32; Sapulpa Historical Society, *Sapulpa, Ok 74066,* pp. 120-121.

[62]Frazier, "History of Glenn Pool," pp. 63-64; Gregory, "How Glenn Pool Oil Turned to Gold for Tulsa," p. 32; Sapulpa Historical Society, *Sapulpa, Ok 74066,* pp. 120-121.

[63]Forbes, "Origin and Early Development of the Oil Industry in Oklahoma," pp. 95-96.

[64]Frazier, "History of Glenn Pool," pp. 65, 93; Forbes, "The

Orgin and Early Development of the Oil Industry in Oklahoma," pp. 86-87.

[65]Frazier, "History of Glenn Pool," p. 65.

[66]*Ibid.*, pp. 65, 93: "Red Fork-Kiefer Map".

[67]Ellen Sue Blakey *et. al.*, The *Tulsa's Spirit* (Tulsa: Continental Heritage Press, n.d.), pp. 107-112; Franks, *The Oklahoma Petroleum Industry*, p. 95.

CHAPTER NINE
Tulsa: Oil Boom Town

[1]Roberson, "City of the Plains, pp. 1-2, 10; Shirk, *Oklahoma Place Names*, p. 209; Angie Debo, *Tulsa: From Creek Town to Oil Capital* (Norman: University of Oklahoma Press, 1943), pp. 3-24; Gregory, "How Glenn Pool Oil Turned to Gold for Tulsa," p. 6.

[2]Roberson, "City of the Plains," pp. 40-44, 49-50; Gregory, "How Glenn Pool Oil Turned to Gold for Tulsa," p. 6; Dunn, *Tulsa's Magic Roots*, p. 226.

[3]Frazier, "History of the Glennpool Oil Field," p. 45.

[4]Blakey *et. al., The Tulsa Spirit*, pp. 27,40; Robinson, "City on the Plains," p. 83; Dunn, *Tulsa's Magic Roots*, pp. 220-226

[5]Blakey *et. al., The Tulsa Spirit*, pp. 27,40; Robinson, "City on the Plains," p. 83; Dunn, *Tulsa's Magic Roots*, pp. 220-226; Gregory, "How Glenn Pool Oil Turned to Gold for Tulsa," p. 15.

[6]Blakey *et. al., The Tulsa Spirit*, pp. 27,40; Robinson, "City on the Plains," p. 83; Dunn, *Tulsa's Magic Roots*, pp. 220-226; Gregory, "How Glenn Pool Oil Turned to Gold for Tulsa," p. 15.

[7]Blakey *et. al., The Tulsa Spirit*, p. 27; Robinson, "City on the Plains," p. 83.

[8]Franks and Lambert, *Early Oklahoma Oil*, p. 108; Gregory, *Oil in Oklahoma*, p. 7, 14.

[9]Franks, Lambert, and Tyson, *Early Oklahoma Oil*, pp. 107-108; Gregory, *Oil in Oklahoma*, pp 7, 14.

[10]Blakey *et. al., The Tulsa Spirit*, p. 180; Dunn, *Tulsa's Magic Roots*, pp. 274-276.

[11]Blakey *et. al., The Tulsa Spirit*, p. 180.

[12]*Ibid.*

[13]Gregory, "When Glenn Pool Oil Turned to Gold for Tulsa," p. 16.

[14]*Ibid.*

[15]*Ibid.*

[16]Dunn, *Tulsa's Magic Roots*, p. 275.

[17]*Ibid.*, pp. 275-276.

[18]*Ibid.*, p. 276.

[19]*Ibid.*

[20]*Ibid.*

[21]*Ibid.*

[22]*Ibid.*, pp. 276-277.

[23]*Ibid.*, p. 277.

[24]*Ibid.*

[25]*Ibid.*, pp. 277-278.

[26]*Ibid.*

[27]Gregory, *Oil in Oklahoma*, pp. 5-10.

[28]*Ibid.*, pp. 6-7; Gregory, "When Glenn Pool Oil Turned to Gold for Tulsa," p. 35; Franks and Lambert, *Early Oklahoma Oil*, pp. 108-109.

[29]Gregory, "When Glenn Pool Oil Turned to Gold for Tulsa," pp. 35-37; Franks and Lambert, *Early Oklahoma Oil*, pp. 108-109; Gregory, *Oil in Oklahoma*, pp. 6-7.

[30]Gregory, "When Glenn Pool Oil Turned to Gold for Tulsa," pp. 35-37; Franks and Lambert, *Early Oklahoma Oil*, pp. 108-109; Gregory, *Oil in Oklahoma*, pp. 6-7.

[31]Gregory, "When Glenn Pool Oil Turned to Gold for Tulsa," pp. 35-37; Franks and Lambert, *Early Oklahoma Oil*, pp. 108-109; Gregory, *Oil in Oklahoma*, pp. 6-7.

[32]Franks and Lambert, *Early Oklahoma Oil*, pp. 108-109; Gregory, *Oil in Oklahoma*, pp. 6-7.

[33]Franks and Lambert, *Early Oklahoma Oil*, pp. 108-109; Gregory, *Oil in Oklahoma*, pp. 6-7.

[34]Franks and Lambert, *Early Oklahoma Oil*, pp. 108-109; Gregory, *Oil in Oklahoma*, pp. 6-7.

[35]Franks and Lambert, *Early Oklahoma Oil*, p. 109; Gregory, *Oil in Oklahoma*, p. 10.

[36]Franks and Lambert, *Early Oklahoma Oil*, p. 109; Gregory, *Oil in Oklahoma*, p. 10.

[37]Franks and Lambert, *Early Oklahoma Oil*, p. 109; Gregory, *Oil in Oklahoma*, p. 10; Debo, *Tulsa*, p. 99.

[38]Franks and Lambert, "Oklahoma Oil Boom Towns"; Forbes, "Southwestern Oil Boom Towns," pp. 399-400.

CHAPTER TEN
Tulsa: Oil Capital of the World

[1]Franks and Lambert, "Oklahoma Oil Boom Towns"; Forbes, "Southwestern Oil Boom Towns," p. 399.

[2]Dunn, *Tulsa's Magic Roots*, pp. 220-221.

[3]*Ibid.*

[4]Gregory, "When Glenn Pool Oil Turned to Gold for Tulsa," pp. 12-14; Blakey *et. al., Tulsa Spirit*, pp. 11, 26-27.

[5]Blakey *et. al., Tulsa Spirit*, pp. 26-27; Gregory, "When Glenn Pool Oil Turned to Gold for Tulsa, pp. 26-27.

[6]Gregory, "When Glenn Pool Oil Turned to Gold for Tulsa," p. 32.

[7]Franks and Lambert, "Oklahoma Oil Boom Towns"; Forbes, "Southwestern Oil Boom Towns," p. 399.

[8]Morris, Goins, and McReynolds, *Historical Atlas of Oklahoma*, map 64; Franks and Lambert, "Oklahoma Oil Boom Towns"; Forbes, "Southwestern Oil Boom Towns," p. 399; Franks, *The Oklahoma Petroleum Industry*, p. 37.

[9]Roberson, "City of the Plains," pp. 82-83; Carl N. Tyson, James H. Thomas, and Odie B. Faulk, *The McMan: The Lives of Robert M. McFarlin and James A. Chapman* (Norman: University of Oklahoma Press, 1977), p. 116; Blakey, et. al, *The Tulsa Spirit*, pp. 27, 42; Franks and Lambert, *Early Oklahoma Oil*, p. 108.

[10]Debo, *Tulsa*, p. 98.

[11]Roberson, "City of the Plains," pp. 82-83; Blakey *et. al., The Tulsa Spirit*, pp. 27, 42; Franks and Lambert, *Early Oklahoma Oil*, p. 108; Gregory, *Oil in Oklahoma*, p. 10.

[12]Roberson, "City of the Plains," p. 91; Blakey *et. al., The Tulsa Spirit*, pp. 25, 39-40; Gregory, "How Glenn Pool Oil Turned to Gold for Tulsa," p. 12.

[13]Roberson, "City of the Plains," p. 91; Blakey *et. al., The Tulsa Spirit,* pp. 25-26, 38; Gregory, "How Glenn Pool Oil Turned to Gold for Tulsa," p. 12.

[14]Roberson, "City of the Plains," pp. 65-80; Blakey *et. al., The Tulsa Spirit,* p. 42.

[15]Roberson, "City of the Plains," p. 81; Galbreath, *Glenn Pool,* pp. 104-106; Blakey *et. al., The Tulsa Spirit,* p. 43; James, *The Texaco Story,* p. 34.

[16]Roberson, "City of the Plains," p. 81; Blakey *et. al., The*

[17]Franks, *The Oklahoma Petroleum Industry,* p. 212; Blakey *et. al., The Tulsa Spirit,* p. 27.

[18]Franks, *The Oklahoma Petroleum Industry,* p. 212-213.

[19]*Ibid.*

[20]*Ibid.,* pp. 213-215.

[21]Gregory, "When Glenn Pool Oil Turned to Gold for Tulsa," p. 35; Franks and Lambert, *Early Oklahoma Oil,* pp. 108-109; Gregory, *Oil in Oklahoma,* p. 9; Dunn, *Tulsa's Magic Roots,* p. 226.

[22]Gregory, *Oil in Oklahoma,* p. 45; Roberta Louise Ironside, *An Adventure Called Skelly: A History of Skelly Oil Company Through Fifty Years, 1919-1969* (New York: Appleton-Century-Crofts, 1970), pp. 1-15.

[23]Gregory, *Oil in Oklahoma,* p. 45; Ironside, *An Adventure Called Skelly,* pp. 1-15, 25, 43.

[24]Gregory, *Oil in Oklahoma,* p. 47; Ironside, *An Adventure Called Skelly,* p. 15.

[25]Gregory, *Oil in Oklahoma,* pp. 46, 48.

[26]Franks, *The Oklahoma Petroleum Industry,* p. 152.

[27]*Ibid.,* pp. 151-154; Clinton, "The Beginnings of the International Petroleum Exposition and Congress," pp. 480-481.

[28]Franks, *The Oklahoma Petroleum Industry,* pp. 151-156; Franks and Lambert, *Early Oklahoma Oil,* pp. 109-110; Clinton, "The Beginnings of the International Petroleum Exposition and Congress," pp. 480-481, 481n. Among those serving on the various committies were: L. D. Armstrong, A. L. Beekly, Carl Blackman, A. V. Bourque, O. V. Borden, A. F. Bourne, F. W. Bryant, D. E. Buchanan, John Campion, Fred S. Clinton, O. L. Cordell, E. H. Cornelius, A. B. C. Dague, W. E. Espy, C. T. Everett, Charles F. Farren, and E. H. Filley. Others included: T. M. Pariss, N. R. Graham, R. D. Gwynne, J. Burr Gibbons, Frank Glasscock, E. Bee Guthrie, R. L. Ginter, H. H. Goddard, J. H. Gardner, J. M. Hayner, Alf G. Heggem, T. J. Hartman, A W. Hurley, Frank Hinderliter, Summers Hardy, Richard Hughes, A. C. Holmen, W. R. Hamilton, R. P. Humes, L. B. Jackson, and Cornelius Kroll. In addition other committees were staffed by L. E. Kennedy, I. G. Long, W. L. Lewis, C. M. Lemason, J. J. McGraw, W. A. Melton, P. M. Miskel, Charles Meyers, Everett Manning, J. H. McBirney, D. W. Moffitt, G. L. Matson, R. L. McFarland, T. F. Mayer, C. M. Murray, H. E. McElroy, J. S. McKelvey, Hollis P. Porter, Asa E. Ramsey, I. G. Rosser, E. A. Richards, Ralph C. Riley, Harold E. Roe, W. R. Richie, Harry Smith, J. A. Sartori, E. T. Tucker, J. A. Udden, W. A. Vanderver, M. M. Valerius, E. H. Wiet, Allan Whiteside, J. S. Warren, W. M. Welch, J. E. Minger, and John Zink.

[29]Franks, *The Oklahoma Petroleum Industry,* pp. 155-156; Franks and Lambert, *Early Oklahoma Oil,* pp. 109-110; Clinton, "The Beginnings of the International Petroleum Exposition and Congress," p. 480.

[30]Franks, *The Oklahoma Petroleum Industry,* pp. 155-160; Franks and Lambert, *Early Oklahoma Oil,* pp. 109-110; Clinton, "The Beginnings of the International Petroleum Exposition and Congress," p. 480.

[31]Franks, *The Oklahoma Petroleum Industry,* pp. 159-160; Gregory, *Oil in Oklahoma,* p. 51.

[32]Franks and Lambert, *Early Oklahoma Oil,* p. 110; Franks, *The Oklahoma Petroleum Industry,* pp. 159-160, 224-225.

[33]Franks, *The Oklahoma Petroleum Industry,* pp. 151, 161; Franks and Lambert, *Early Oklahoma Oil,* pp. 109-110; Debo, *Tulsa,* 109.

[34]Franks, *The Oklahoma Petroleum Industry,* pp. 151, 161; Franks and Lambert, *Early Oklahoma Oil,* pp. 109-110; Debo, *Tulsa,* 109.

[35]Guy Logsdon, *The University of Tulsa: A History, 1882-1972* (Norman: Unversity of Oklahoma Press, 1977), pp. 14-142; Debo, *Tulsa,* p. 106; Blakey *et. al., Tulsa Spirit,* pp. 28, 123.

[36]Debo, *Tulsa,* p. 106; Blakey *et. al., Tulsa Spirit,* pp. 28, 123; Franks, *The Oklahoma Petroleum Industry,* pp. 196-198; Gregory, "When Glenn Pool Oil Turned to Gold for Tulsa," p. 25.

[37]Debo, *Tulsa,* p. 106; Blakey *et. al., Tulsa Spirit,* pp. 28, 123; Franks, *The Oklahoma Petroleum Industry,* pp. 196-198; Logsdon, The University of Tulsa, pp. 163n, 217n, 222-223n, 258-259n, 298-299, 306.

[38]Franks, *The Oklahoma Petroleum Industry,* p. 200; Blakey *et. al., The Tulsa Spirit,* pp. 35-36, 108-112; Litton, *History of Oklahoma,* Vol. II, p. 193.

[39]Blakey *et. al., The Tulsa Spirit,* pp. 35-36, 108-112.

CHAPTER ELEVEN
Glenn Pool's Legacy

[1]Franks, *The Oklahoma Petroleum Industry,* pp. 45-46; Franks, Lambert, Tyson, *Early Oklahoma Oil,* pp. 39-40; Frazier, "History of the Glennpool Oil Field," p. 56; Gregory, "How Glenn Pool Oil Turned to Gold for Tulsa," p. 39; American Petroleum Institute, *Petroleum Facts and Figures, Ninth Edition, 1950,* p. 156; Williamson, Andreano, Daum, and Klose, *The American Petroleum Industry,* Vol. II, p. 18; Rister, *Oil!,* p. viii; Franks, Lambert, and Tyson, *Early Oklahoma Oil,* p. 3.

[2]Rister, *Oil!,* pp. viii, 410-411; Franks, Lambert, and Tyson, *Early Oklahoma Oil,* p. 3; American Petroleum Institute, *Petroleum Facts and Figures, Ninth Edition, 1950,* pp. 152-153; *Rister!,* pp. 410-411.

[3]American Petroleum Institute, *Petroleum Facts and Figures, Ninth Edition,* 1950, pp. 144-147.

[4]Franks, *The Oklahoma Petroleum Industry,* pp. 45-46; Franks, Lambert, Tyson, *Early Oklahoma Oil,* pp. 39-40; Frazier, "History of the Glennpool Oil Field," p. 56; Gregory, "How Glenn Pool Oil Turned to Gold for Tulsa," p. 39; American Petroleum Institute, *Petroleum Facts and Figures, Ninth Edition, 1950,* p. 156.

[5]American Petroleum Institute, *Facts and Figures, Ninth Edition, 1950,* pp. 152-153; Rister, *Oil!,* pp. 410-411.

[6]Williamson, Andreano, Daum, and Klose, *The American Petroleum Industry,* Vol. II, pp. 22-23.

[7]James, *The Texaco Story,* pp. 16-19.

[8]*Ibid.,* pp. 19-21.

[9]*Ibid.,* pp. 21-25.

[10]*Ibid.;* Williamson, Andreano, Daum, and Klose, *The American Petroleum Industry,* Vol. II, pp. 170-174.

[11]James, *The Texaco Story,* pp. 21-25; Williamson, Andreano, Daum, and Klose, *The American Petroleum Industry,* Vol. II, pp. 170-174, 192-195.

[12]James, *The Texaco Story,* pp. 21-27.

[13]*Ibid.*, pp. 18, 32-33.

[14]Thompson, *Gulf Since Spindletop,* pp. 14, 17-23.

[15]*Ibid.*, pp. 23-27.

[16]Franks, *The Oklahoma Petroleum Industry*, p. 18; James, *The Texaco Story,* p. 32.

[17]Franks, *The Oklahoma Petroleum Industry*, p. 18; James, *The Texaco Story,* p. 32.

[18]Franks, *The Oklahoma Petroleum Industry*, p. 18; James, *The Texaco Story,* p. 32; Roberson, "City of the Plains," pp. 65-80; Blakey *et. al., The Tulsa Spirit,* p. 42.

[19]Franks, *The Oklahoma Petroleum Industry*, p. 42; Charles E. Bowles, "Oklahoma Petroleum—An Industrial Survey," *Oklahoma Geological Survey Bulletin No. 40,* Vol. I, *Oil and Gas in Oklahoma, p.* 88.

[20]Bowles, "Oklahoma Petroleum—An Industrial Survey," pp. 88-89; Franks, *The Oklahoma Petroleum Industry*, p. 46.

[21]Bowles, "Oklahoma Petroleum—An Industrial Survey," pp. 88-89; Franks, *The Oklahoma Petroleum Industry*, p. 46.

[22]Bowles, "Oklahoma Petroleum—An Industrial Survey," pp. 83-84; Boone, *The Petroleum Dictionary,* pp. 83, 206; Lambert, "Natural Gas," Morris, ed., *Drill Bits, Picks, and Shovels: A History of Mineral,* p. 57.

[23]Lambert, "Natural Gas," p. 57.

[24]American Petroleum Institute, *Petroleum Facts and Figures, Ninth Edition, 1950,* p. 202; Bowles, "Oklahoma Petroleum—An Industrial Survey," p. 87.

[25]American Petroleum Institute, *Petroleum Facts and Figures, Ninth Edition, 1950,* p. 202; Bowles, "Oklahoma Petroleum—An Industrial Survey," p. 87.

[26]Boone, *Petroleum Dictionary,* pp. 45,95; Bowles, "Oklahoma Petroleum—An Industrial Survey," pp. 95-99.

[27]Bowles, "Oklahoma Petroleum—An Industrial Survey," pp. 95-99.

[28]Lambert, "Natural Gas," p. 54; Ed Wheeler, "Oklahoma Natural Gas Company: A Profile," *Gasette,* Vol. 33, No. 3 (October/November, 1974), p. 2.

[29]Ferrell, "The Oklahoma Natural Gas Company," pp. 9-10.

[30]*Ibid.*, pp. 10-12.

[31]Wheeler, "Oklahoma Natural Gas Company," pp. 2-3; Lambert, "Natural Gas," p. 54; Ferrell, "The Oklahoma Natural Gas Company," pp. 12-13.

[32]Wheeler, "Oklahoma Natural Gas Company," pp. 2-3; Lambert, "Natural Gas," p. 54; Ferrell, "The Oklahoma Natural Gas Company," pp. 12-13.

[33]Wheeler, "Oklahoma Natural Gas Company," pp. 2-3; Lambert, "Natural Gas," p. 54; Ferrell, "The Oklahoma Natural Gas Company, p. 14-15.

[34]Wheeler, "Oklahoma Natural Gas Company," pp. 2-3; Lambert, "Natural Gas," p. 54; Ferrell, "The Oklahoma Natural Gas Company, p. 14-15.

[35]Lambert, "Natural Gas," p. 54; Wheeler, "Oklahoma Natural Gas Company," pp. 2-3; "Oklahoma Natural Gas Company: A Brief History," Cleveland, Red Fork, and Glenn Pool Papers, Oklahoma Petroleum Collection, Archives, Oklahoma Heritage Association, Oklahoma City, Oklahoma; Bowles, "Oklahoma Petroleum—An Industrial Survey," p. 87; American Petroleum Institute, *Petroleum Facts and Figures, Ninth Edition, 1950,* pp. 178-179.

[36]American Petroleum Institute, *Petroleum Facts and Figures, Ninth Edition, 1950,* pp. 178-179; Bowles, "Oklahoma Petroleum—An Industrial Survey," p. 87.

[37]Franks, *The Oklahoma Petroleum Industry*, pp. 36-46, 226-227; Gregory, *Oil in Oklahoma*, pp. 45-51, 58-69.

[38]David R. Milsten, *Thomas Gilcrease* (San Antonio, Texas: The Naylor Company, 1969), pp. 11-55; Gregory, *Oil in Oklahoma*, pp. 60-61; Gregory, "When Glenn Pool Oil Turned to Gold for Tulsa," p. 32.

[39]Milsten, *Thomas Gilcrease,* pp. 49-79; Gregory, "When Glenn Pool Oil Turned to Gold for Tulsa," p. 32; Gregory, Oil in Oklahoma, pp. 61-63.

[40]Gregory, *Oil in Oklahoma*, pp. 61-69; Franks, *The Oklahoma Petroleum Industry,* pp. 226-227.

[41]"Alfred E. Aaronson," Archives, Oklahoma Heritage Association, Oklahoma City, Oklahoma; Franks, *The Oklahoma Petroleum Industry,* pp. 226-227; Gregory, *Oil in Oklahoma*, p. 65.

[42]Gregory, *Oil in Oklahoma*, p. 65; Franks, *The Oklahoma Petroleum Industry,* pp. 226-227.

[43]Gregory, *Oil in Oklahoma*, pp. 65-69; Franks, *The Oklahoma Petroleum Industry,* pp. 226-227; Milsten, *Thomas Gilcrease,* p. 387.

[44]Gregory, *Oil in Oklahoma*, pp. 65-69; Franks, *The Oklahoma Petroleum Industry,* pp. 226-227; Milsten, *Thomas Gilcrease,* p. 387.

[45]Franks, *The Oklahoma Petroleum Industry*, p. 46.

BIBLIOGRAPHY

Manuscript Material

"Alfred E. Aaronson." Archives. Oklahoma Heritage Association, Oklahoma City, Oklahoma.

Lambert, Paul F. and Kenny A. Franks. "Oklahoma Oil Boom Towns." Oklahoma Petroleum Collection. Archives. Oklahoma Heritage Association, Oklahoma City, Oklahoma.

"Oklahoma Natural Gas Company: A Brief History." Cleveland, Red Fork, and Glenn Pool Papers. Oklahoma Petroleum Collection. Archives. Oklahoma Heritage Association, Oklahoma City, Oklahoma.

Government Records

Department of Economic and Community Affairs, Oklahoma State Data Center. "1980 Census Data: Population and Household Characteristics," United States Documents Regional Depository, Oklahoma Department of Libraries, Oklahoma City, Oklahoma.

Oklahoma Corporation Commission, Oil & Gas Conservation Division. "Annual Summary by Quarter of Liquids & Gas Runs by Counties, 1980," Oklahoma Publications Clearinghouse, Oklahoma Department of Libraries, Oklahoma City, Oklahoma.

Oklahoma Corporation Commission, Oil & Gas Conservation Division. "January, 1983, Liquids and Gas Runs by Counties," Oklahoma Publications Clearinghouse, Oklahoma Department of Libraries, Oklahoma City, Oklahoma.

Interviews

"Indian-Pioneer History." Grant Foreman, ed., 113 vols., Indian Archives. Oklahoma Historical Society. Dillon, Jack, Vol. III, p. 41

Cleveland, Red Fork, and Glenn Pool Papers. Oklahoma Petroleum Collection. Archives. Oklahoma Heritage Association, Oklahoma City, Oklahoma.

Hubbard, James W. September 27-28, 1983.

Jordan, Robert O. September 26, 1983.

Lumsford, Robert L., III. September 26, 1983.

Theses and Dissertations

Ferrell, Robert A. "The Oklahoma Natural Gas Company: A Study in Industrial Geography." M.A. Thesis, University of Oklahoma, 1964.

Forbes, Charles G. "The Origin and Early Development of the Oil Industry in Oklahoma." Ph.D. Dissertation, University of Oklahoma, 1939.

Frazier, J. Vere, Jr. "History of the Glenn Pool Oil Field." M.A. Thesis, University of Oklahoma, 1951.

Roberson, Glen R. "City of the Plains: The History of Tulsa, Oklahoma." Ph.D. Dissertation, Oklahoma State University, 1977.

Newspapers

Sunday Oklahoman. Oklahoma City, Oklahoma.

The Triangle. Cleveland, Oklahoma.

Tulsa Democrat. Tulsa, Oklahoma.

Maps

"Cleveland, Boston, Lauderdale, and Maramec Oil and Gas Pools." *Petroleum and Natural Gas in Oklahoma, Oklahoma Geological Survey Bulletin No. 19,* Part II (April, 1917), Plate XXXV.

"Creek County Map No. 2." Oklahoma Department of Transportation. Oklahoma City, Oklahoma.

"Tulsa County Map." Oklahoma Department of Transportation. Oklahoma City, Oklahoma.

"Pawnee County Map." Oklahoma Department of Transportation. Oklahoma City, Oklahoma.

"Red Fork-Kiefer Map." *Oklahoma Geological Survey Bulletin No. 19, Part II* (April, 1917), Plate XVII.

Published Material

American Petroleum Institute, *Petroleum Facts and Figures, 1950.* New York: American Petroleum Institute, 1951.

Blakey, Ellen Sue *et. al. The Tulsa Spirit.* Tulsa: Continental Heritage Press, n.d.

Boone, Lalia Phipps. *The Petroleum Dictionary.* Norman: University of Oklahoma Press, 1952.

Bowels, Charles E. "Oklahoma Petroleum—An Industrial Survey." *Oklahoma Geological Survey Bulletin No. 40,* Vol. I, *Oil and Gas in Oklahoma,* pp. 79-97.

Carpenter, Everett. "The Morrison Field, Pawnee County Oklahoma," *Oil and Gas in Oklahoma, Oklahoma Geological Survey Bulletin No. 40* (July, 1930), pp. 184-195.

Clark, J. Stanley. *The Oil Century: From the Drake Well to the Conservation Era*. Norman: University of Oklahoma Press, 1958.

Clinton, Fred S. "The Beginning of the International Petroleum Exposition and Congress." *The Chronicles of Oklahoma*, Vol. XXVI, No. 4 (Winter, 1948-1949), pp. 479-488.

__________. "First Oil and Gas Well in Tulsa County," *The Chronicles of Oklahoma*, Vol. XXX, No. 3 (Autumn, 1952), pp. 312-332.

Colcord, Charles Francis. *The Autobiography of Charles Francis Colcord, 1859-1934*. Privately printed, 1970.

Connelly, W. L. *The Oil Business As I Saw It: Half a Century with Sinclair.* Norman: University of Oklahoma Press, 1954.

Corden, Seth K. Corden and W. B. Richards, comps. *The Oklahoma Red Book.* 2 vols., Oklahoma City: Democrat Printing Company, 1912.

Craighead, David. "Glenn Pool: A Forest of Derricks," *Oklahoma Today*, Vol. XVI, No. 1 (Winter 1965-1966), pp. 9-11

Debo, Angie. *Tulsa: From Creek Town to Oil Capital.* Norman: University of Oklahoma Press, 1943.

Dille, Glen S. "Economic Geology: Oil and Gas in Tulsa County," in Malcolm C. Oakes, ed., *Geological and Mineral Resources of Tulsa County Oklahoma, Oklahoma Geological Survey Bulletin No. 69* (1952), pp. 104-125.

Forbes, Gerald. "Southwestern Oil Boom Towns," *The Chronicles of Oklahoma*, Vol. XVII, No. 4 (December, 1939), pp. 393-400.

Franks, Kenny A. *The Oklahoma Petroleum Industry.* Norman: University of Oklahoma Press, 1980.

Franks, Kenny A., Paul F. Lambert, and Carl N. Tyson. *Early Oklahoma Oil: A Photographic History, 1859-1936.* College Station: Texas A&M University Press, 1981.

Galbreath, Frank. *Glenn Pool...and a Little Oil Town of Yesteryear.* Privately printed, 1978.

Getty, J. Paul. *The History of the Oil Business of George F. and J. Paul Getty from 1903 to 1939*. Privately printed, 1941.

__________. *As I See It: The Autobiography of J. Paul Getty.* Englewood Cliffs, New Jersey: Prentice-Hall, Inc., 1976.

__________. *My Life and Fortunes.* New York: Duel, Sloan and Pearce, 1963.

Glasscock, C. B. *Then Came Oil: The Story of The Last Frontier.* New York: The Bobbs-Merrill Company, 1938.

Gregory, Bob. "How Glenn Pool Oil Turned to Gold For Tulsa," *Tulsa Magazine*, Vol. 58, No. 5 (November 20, 1980), pp. 5-39.

Greig, Paul B. *Geology of Pawnee County, Oklahoma, Oklahoma Geological Survey Bulletin No. 83* (1959).

Greene, Frank C. "Pawnee County," *Oil and Gas in Oklahoma, Oklahoma Geological Survey Bulletin No. 40* (July, 1930), pp. 169-185.

Gunther, Max. *The Very, Very Rich and How They Got That Way.*Chicago: Playboy Press, 1972.

Hewins, Ralph. *The Richest American: J. Paul Getty.* New York: E. P. Dutton & Co., Inc., 1960.

Ironside, Roberta Louise. *An Adventure Called Skelly: A History of Skelly Oil Company Through Fifty Years, 1919-1969.* New York: Appleton-Century-Crofts, 1970.

James, Marquis. *The Texaco Story: The First Fifty Years, 1902-1952.* N.p., The Texas Company, 1953.

Kappler, Charles J., comp. *Indian Affairs: Laws and Treaties.* 5 vols., New York: AMS Press Inc., 1971.

Lambert, Paul F. "Natural Gas: Fuel for the Future." John W. Morris, ed., *Drill Bits, Picks, and Shovels: A History of Mineral Resources in Oklahoma.* Oklahoma City: Oklahoma Historical Society, 1982, pp. 49-81.

Logsdon, Guy. *The University of Tulsa: A History, 1882-1972.* Norman: Unversity of Oklahoma Press, 1977.

Litton, Gaston. *History of Oklahoma at the Golden Anniversary of Statehood.* 4 vols., New York: Lewis Historical Publishing Company, 1957.

McComas, Walter G. "Sapulpa, The Metropolis of the World's Greatest Oil Museum," *Sturm's Oklahoma Magazine*, Vol. IV, No. 2 [3?] (May, 1908), pp. 24-31.

Mills-Bullard, Bess, comp. "Digest of Oklahoma Oil and Gas Fields," *Oil and Gas in Oklahoma, Oklahoma Geological Survey Bulletin No. 40*, Vol. I (July, 1928), pp. 101-276.

Milsten, David R. *Thomas Gilcrease.* San Antonio, Texas: The Naylor Company, 1969.

Morris, John W., Charles R. Goins, and Edwin C. McReynolds. *Historical Atlas of Oklahoma.* Norman: University of Oklahoma Press, 1976.

Oakes, Malcolm C. "Geography," in Malcolm C. Oakes, ed., *Geological and Mineral Resources of Tulsa County Oklahoma, Oklahoma Geological Survey Bulletin No. 69* (1952), pp. 13-16.

Rister, Carl C. *Oil! Titan of the Southwest.* Norman: University of Oklahoma Press, 1949.

Roberts, Joe Donald. "Ten-Barrel 'Whodunit' at Red Fork," *The Chronicles of Oklahoma,* Vol. LVI, No. 4 (Winter, 1978-1979), pp. 442-449.

Sapulpa Historical Society. *Sapulpa, Ok 73066.* 2 vols., Sapulpa: Sapulpa Historical Society, 1981.

Shirk, George H. *Oklahoma Place Names.* Norman: University of Oklahoma Press, 1966.

Sinclair Oil Corporation. *A Great Name in Oil: Sinclair Through Fifty Years.* New York: McGraw-Hill, Inc., 1966.

Thompson, Craig. *Gulf Since Spindletop: A Human Story of Gulf's First Half-Century.* Privately printed, n.d.

Tyson, Carl N., James H. Thomas, and Odie B. Faulk. *The McMan: The Lives of Robert M. McFarlin and James A. Chapman.* Norman: University of Oklahoma Press, 1977.

Wheeler, Ed. "Oklahoma Natural Gas Company: A Profile." *Gasette,* Vol. 33, No. 3 (October/November, 1974), pp. 1-25.

Williamson, Harold F., Ralph L. Andreano, Arnold R. Daum, and Gilbert Klose. *The American Petroleum Industry.* 2 vols., Evanston, Illinois: Northwestern University Press, 1959-1963.

INDEX

AAPG: *see* American Association of Petroleum Geologists
Aaronson, Alfred E.: 141–142
A. C. F. Gasoline Company: 137
Act for the Protection of the People of the Indian Territory and for other Purposes: 83; *see also* Curtis Act
Adams, H. H.: 10, 14, 17
Adams, Lewis: 10
Adams, Thomas J.: 10
Adams, Wash: 10
Adamson Field: 47
Alcorn Hotel (Tulsa): 108
American Association of Petroleum Geologists: 127
American National Bank (Sapulpa): 88
American Saturday Night: 123
Ames, C. B.: 139–140
Amesville, Louisiana: 133
Amity Gasoline Company: 137
Angelina, Texas: 81
Anthony, S. W.: 106
Appalachia, Oklahoma: 47
A & P Railroad: *see* Atlantic and Pacific Railroad
Arkansas: 72
Arkansas River: 3, 5, 9–10, 14, 17, 26–28, 35–36, 43, 45–47, 64, 79, 83, 108, 111, 119
Arkansas Valley Oil and Gas Company: 138
Arlington Hotel, Cleveland: 34
Armold H. Y.: 81
Arnold, E. M.: 113
Arthur Oil Company: 137
Asphaltus Spring: 27
Associated Oil Company: 73
Associated Petroleum Company: 76
Atchison, Topeka and Santa Fe Railroad: 118
Atkins, Louis: 12
Atkins, R. D.: 17
Atlantic and Pacific Railroad: 3, 5
Atoka, Oklahoma: 81
Atwood, Weston: 62–63
Ault, A. F.: 113
Aylesworth, Allison: 14–15
Aztec Oil Company: 137

Baird, M. L.: 120
Baker farm: 53
Baker lease: 53
Baltimore, Maryland: 117
Bank of Oklahoma (Tulsa): 119
Barde, Fred S.: 17
Barnsdall Oil Company: 137
Barnsdall, Theodore N.: 138–140
Barnes Investment Company: 70
Bartlesville, Oklahoma: 51, 135
Bartlesville Sand: 36, 39
Bartlett, R. H.: 140
Baruer, Robert: 63
Battle Axe Railroad: 76
Baston, Texas: 133
Beacon Building (Tulsa): 122
Beaumont, Texas: 121
Bechtel, W. F.: 47
Beckstrom, Robert C.: 126
Bell, John W.: 51
Benedum, Mike: 9
Berryhill, D. L.: 10
Berryhill, Earl: 69
Berryhill, Joe: 106
Berryhill, Joseph F.: 70
Bigheart, Oklahoma: 115
Big Lime Horizon: 75
Big Pasture: 27
Big Sandy, Texas: 81
Binns, Andy: 22
Blakemore, J. S.: 15, 17
Blackburn East Field: 47
Blackburn Southeast Field: 47
Blackburn South Field: 47
Bland, Arlies: 12, 25
Bland, Davis: 12, 25
Bland, Era: 12, 17, 25
Bland, Hazel
Bland, John C. W.: 9–26
Bland, Lee: 17
Bland, Louise: 17
Bland, Owen: 12, 17, 25
Bland, Paul: 17
Bland, Sue A.: 10–26
Bland, Vera: 12, 17, 25
Bonaker, Ed: 18, 76
Born, Sidney: 126
Boston Avenue Methodist Church (Tulsa): 113
Boston Field: 43
Bourque, A. V.: 124
Boviard Supply Company (Sapulpa): 88
Bowden, Oklahoma: 87, 105; *see also* Taneha
Bowery, the: 102, 104
Boyle, Patrick C.: 121
Braden, Glenn T.: 140
Brady Store: 108
Branden, Glenn: 22–23
Breckenridge, Clifton R.: 19
Bridges, Mr. and Mrs. H. E.: 17
Bristow, Oklahoma: 104
Brock No. 1: 70
Brock No. 2: 71
Brock, Eck E.: 62; farm of, 67; lease of, 71
Brown, Frank O.: 14, 17
Brown Frederick V.: 47
Bryan Field: 56
Bryan North Field: 56
Bryan Northeast Field: 56
Bryan Northwest Field: 56
Bryan Southeast Field: 56
Buck, F. J.: 113
Bucket of Blood Saloon (Tulsa): 111
Burgess Horizon: 75
Burnett, Bates: 63
Buttram, Frank: 43
Byrd, Edward: 5
Byrne, George: 115
Byrum Store: 108

Cache Creek: 27
Caddo, Oklahoma: 27
Callery, James D.: 135
Canada: 117
Canadian River: 81
Carson, John: 17
Carson lease: 53
Carter Brothers Oil Company: 45
Carter County, Oklahoma: 56
Carter Oil Company: 126
Casey Field: 45

INDEX

Casey North Field: 45
Cedar Creek: 30, 35
Centerville, Iowa: 12
Chambers, Oklahoma: 81
Chandler, Oklahoma: 27
Chapman Building (Tulsa): 126
Chapman, James A.: 118–119, 122, 126
Chas. Colclezer No. 1: 45
Chelsea, Oklahoma: 5
Cherokee Indians: 35
Cherokee Nation: 135
Cherokee Outlet: 27–28
Cherryvale, Kansas: 79
Chesley, Frank: 64–66
Chestnut & Smith Corporation: 137
Cheyenne and Arapaho Reservation: 27
Chicago, Illinois: 81, 117, 140
Chickasaw Oil Company: 5
Chief Mining Inspector: 85
Chisholm Trail: 12
Chronicles of Oklahoma, The: 10
Cimarron River: 5, 28, 64
Cincinnati, Ohio: 117
Cleveland Field: 27–61
Cleveland, Grover: 28
Cleveland, Oklahoma: 27–43
Cleveland Oil, Gas and Manufacturing Company: 33
Cleveland Sand: 36
Clinton allotment: 81
Clinton, Fred S.: 10–26, 122
Clinton, Lee: 17
Club Hotel (Tulsa): 108
"Coal Oil Johnny" (special train): 111
Coast Oil Company: 114
Cobb, Guy P.: 19
Cody Field: 56
Cody Southwest Field: 56
Coffyville, Kansas: 81
Colcord, Charles F.: 17, 22, 23, 62
Colgate, Oklahoma: 81
Colorado School of Mines: 126
Colonel Robinson's Hotel (Red Fork): 14
Col. Moore Hotel (Tulsa): 108
Comanche, Kiowa and Apache Reservation: 27
Commonwealth Company: 141
Connelly, David F.: 67, 75
Connelly, W. L.: 73, 91–92, 94
Corbray, Elijah: 67
Cordova Hotel (Tulsa): 55
Cosden Building (Tulsa): 115
Cosden, Joshua S.: 109, 114–116, 121
Cosden Pipeline Company: 118, 135
Cosden Refinery (West Tulsa): 121
Cottage Home Hotel (Tulsa): 108
Creek Bank & Trust (Sapulpa): 88
Creek County, Oklahoma: 3, 12, 47, 53, 55
Creek Nation: 3, 27, 35; oil in, 3–26
Creek Oil Company: 66, 67, 71, 76
Creek Treaty of 1833: 5
Creek Treaty of March 1, 1901: 23
Crescent Star South Field: 56
Cripple Creek, Colorado: 104
Cromwell, Oklahoma: 116
Crosbie & Gillespie: 37
Crosby, Elizabeth: 62–63
Crossman Brothers: 10–11
Crossman, Luther: 11–12
Crossman, Perry: 11–12, 14, 17
Cruce, Lee: 114
Cudahy, Michael: 5, 135
Cullinan, Joseph S.: 133
Curtis Act: 8–9, 83
Cushing, Oklahoma: 27, 118, 142
Cushing-Drumright Field: 53

Daily Independent, The: 113
Darden, Robert M.: 5
Dave Shultz Dry Goods (Tulsa): 108
David, Samuel C.: 10
Davis, Sue A.: 12
Davis, W. T.: 12
Dawes Act: 7–8, 28
Dawes Commission: 14–15, 19, 23–24, 208
Delaware: 123
DeGolyer, E. L.: 126
Delta, Texas: 81
Devonian Field: 47
Devonian Oil Company: 47
Diamond Refining Company: 137
Dickerson Lumber Company (Tulsa): 108
Dillon, Jack: 104
Doan, Murray M.: 113
Dodd, Roy: 63–65
Dodge City, Kansas: 12
Dog Center, Oklahoma: 36
Donahue, Charley: 18
Donahue Field: 47; *see also* Masham Field
Donahue, J. L.: 47
Drake, Edwin L.: 5
Duck Creek: 12, 105
Dunn, J. M.: 10
Duquesne Petroleum Company: 45
Durant, Edith: 70
D. W. Franchot Company: 135

Eagle Gasoline Company: 137
East Texas Field: 55
Eastern Oil Company: 70
Edgar Brothers' Building (Tulsa): 108
Edmond, Oklahoma: 62
El Dorado, Kansas: 123
Eleventh Street Bridge (Tulsa): 26, 119–120
Elliot Building (Tulsa): 122
Elred, J. C.: 70
Empire Pipeline Company: 118, 135
English, A. Z.: 16
Ethel Oil Company: 114
Evans, J. H.: 118
Exchange National Bank (Tulsa): 119

Farmers and Merchants Bank (Sapulpa): 88
Farmers National Bank (Tulsa): 119
Faucett, H. W.: 5
Fellows, W. J.: 30, 32
Ferguson, Thompson B.: 34, 36
Fife, Timmie: 10
First Christian Church (Tulsa): 108
First Methodist Church (Tulsa): 108
First National Bank (Sapulpa): 88
First National Bank (Tulsa): 114
Fite, F. B.: 14, 15, 17
Flinn, William: 135

INDEX

Flynn, Barney: 106
Flynn, Dennis T.: 139–140
Fort Sill, Oklahoma: 27
Forsythe, C. M.: 17
Fortuna Oil Company: 43
Fort Worth, Texas: 89
Fosburgh, L. L.: 88
Four Gasoline Company: 137
Franchot, D. W.: 101
Franchot Oil Company: 39, 96–97; *see also* D. W. Franchot Company
Franklin County, Texas: 81
Franklin, Wirt: 123
Fred Scott Dry Goods (Tulsa): 108
Frisco Railroad: *see* St. Louis and San Francisco Railroad

Gaberdiel, Bob: 96–98
Galbreath and Company: 73, 75
Galbreath, Frank: 75, 84
Galbreath Gas Company: 83, 138–139
Galbreath, Howard: 91–92, 94
Galbreath Oil and Gas Company: 70
Galbreath, Robert A.: 17, 22–23, 62–71, 75, 83, 138–139
Galey, John: 138
Gardner, J. H.: 123
Garrison, Texas: 132
Garr Field: 47
George F. Getty, Inc.: 55
George L. Miller No. 1: 43
George L. Miller No. 3: 43
Getty, George F.: 3, 47, 51, 53, 55
Getty Oil Company: 3, 51, 55, 73, 122–123
Getty, J. Paul: 47, 51–55, 122
Getty, Sarah MacPherson Risher: 51
Gilcrease Oil Company: 141
Gilcrease, Thomas: 141–142; lease, 70
Gill's Blacksmith Shop (Tulsa): 108
Gladstone Hotel (Sapulpa): 18
Glenn, Elna: 69
Glenn Gas Company: 137
Glenn, Gracie: 67
Glenn Horizon: 75
Glenn, Ida: 72; allotment, 62–63; farm, 63; lease, 63
Glenn, Mable: 67
Glenn No. 1: 83
Glenn Pool: 5, 22, 26, 57, 62–109, 112, 114, 117–120, 132–142; development of, 73–86
Glenpool: 87, 106–107
Glenn Sand: 75
Gotebo, Oklahoma: 27
Grand River: 3
Grand Saline: 5
Granite, Oklahoma: 27
Grant: B. W.: 113
Great Lakes: 81
Greater Osage Field: 138
Greer County, Oklahoma: 27
Greiner, A. J.: 27
Guffey, James M.: 138
Gulf Coast: 135
Gulf Oil Company: 81, 107, 135
Gulf Pipeline Company: 76, 98, 118, 135
Gulf Pipeline Company of Texas: 81
Gulf Refining Company: 81, 135
Gulf Refining Company of Texas: 135
Guthrie, Oklahoma: 27, 63, 83
Guthrie, William S.: 62
G. W. Besser No. 1: 47
Gypsy: 81
Gypsy Oil Company: 43, 81, 137

Haggem, Alf G.: 123
Hagler, Don: 26
Hagler, J. D.: 120
Hayner, J. M.: 123
Half Circle S Ranch: 12
H. C. Hall and Company store (Red Fork): 5
Hall, J. B.: 17
Hall, J. M.: 112
Hall, J. W.: 17
Hallett Field: 45
Hallett Northwest Field: 45
H. M. Byllesby Company: 140
Hardin County, Texas: 81
Hardridge, Eli E.: 10
Hartman, T. J.: 123
Harvey House Restaurant (Sapulpa): 89
Harwell, E. W.: 126
Harwell Gymnasium (Tulsa): 126
Haskell, Charles N.: 85
Healdton, Oklahoma: 142
Healdton-Hewitt Field: 123
Hennessey, Oklahoma: 62
Henryetta, Oklahoma: 81
Henry Kendall College: 125; *see also* University of Tulsa
Herbert, Oklahoma: 28, *see also* Cleveland
Herbert, Willis H.: 28
Heydrick, James A.: 9–26
Hill, J. H.: 14, 17
Hitchcock, Ethan Allen: 12, 19, 24–25
Hogan's Livery (Tulsa): 108
Holden, William: 123
Holland, C. L.: 113
Hominy Horizon: 75
Hominy Sand: 47
Hopkins County, Texas: 81
Hotel Tulsa: 3, 109
Howard Park (Tulsa): 62
Hughes County, Oklahoma: 5
Hughes, Robert: 17
Hughes, Thomas: 17
Hull, Arthur: 113
Humble, Texas: 81
Hunt, O. D.: 114
Hutchinson, L. L.: 53

Ida Glenn No. 1: 62–69, 71, 87, 108, 120, 135
Independence, Kansas: 135
Indian Republican: 108
Indian Territory Field: 135; *see also* Glenn Pool
Indianola Oil and Gas Company: 25
International Petroleum Exposition: 123–125
IPE: *see* International Petroleum Exposition
Iowa: 27
IX Ranch: 16

Jack Archer's Store (Tulsa): 108
Jackson, L. B.: 123
Jane Perryman's Boarding House (Tulsa): 108
Jenks, Oklahoma: 79–87, 106–107
Jenks, Elmer E.: 106
Jennings Field: 45

INDEX

Jennings, Louisiana: 133
J. M. Hall and Company store (Red Fork): 5
J. M. Hall Store (Tulsa): 108
J. M. Morrow Drugs (Tulsa): 108
John Mabee Hall for Men (Tulsa): 126
John Rogers' College of Law Building (Tulsa): 126
John S. Wick Company: 10, 24
Jones, C. G.: 22–23, 62
Joplin, Missouri: 10
Jordan, Robert O.: 30
Josey, R. A.: 108

Kansas: 3, 9, 25, 72, 89, 132
Kansas City, Missouri: 108, 135
Kansas City Times: 17
Kansas Torpedo Company: 33
Kaw Reservation: 27
Kelly-Chandler No. 1: 47
Kellyville, Oklahoma: 140
Kelso Sand: 32
Kemp Hall (Tulsa): 126
Kemp, E. Roger: 126
Kennedy John Q. A.: 10
Kerr, W. A.: 47
Keystone, Oklahoma: 64, 94, 111
Keystone Field: 47, 56
"Keystone Trail" (bootlegger route): 111
Kickapoo Reservation: 27
Kiefer, Anna: 70
Kiefer, Oklahoma: 76, 79, 87, 96, 98, 101, 104–105, 107, 112, 137
Kiefer, Levy R.: 70
Kiefer News: 104
Kiefer, Smith: 96
Kiefer, Stella F.: 70
Kiefer Townsite Company: 101
Knickerbocker, Percy A.: 113
Kolvin, Lete: 53

LaFortune Hall: 126
LaFortune, Joseph A.: 126
Laird, Melvin: 26
Lake, Charles B.: 104
Lamar County, Texas: 81
Land, J. H.: 10
Lauderdale Field: 45
Lauinger, Frank T.: 121
Lauinger, P. C.: 121–122
Lauinger, Phillip, Jr.: 122
Laurel Oil Company: 176
"Law of Capture": 74
Law Enforcement Club (Tulsa): 113–114
Lenior, Texas: 81
Leovy, Frank A.: 81
Lindsey, Lilah D.: 10
Lindsey, Mr. and Mrs. H. D.: 17
Linsky Tailor Shop (Tulsa): 108
Litchfield, Sawyer and Company: 76
Lochapoka Crossing: 3, 5
Longmire, W. M.: 47
Lonnie Price Harness (Tulsa): 108
Lon Stansbery Implements (Tulsa): 108
Lotta Jane Mabee Hall for Women (Tulsa): 126
Louisiana: 72, 132
Louisiana Purchase Exposition and World's Fair: 34
Lower Creeks: 3
Lowery, William: 30
Lufkin, Texas: 81
Lumsford, Robert L., III: 35
Lynch Brothers' General Store (Tulsa): 108
Lynch, R. E.: 113

Mabee Gymnasium Annex (Tulsa): 126
McCloud, Oklahoma: 27
McClure, O. H.: 113, 123
McConnel, P. D.: 51
McFarlin Library (Tulsa): 126
McFarlin, Robert M.: 118–119, 122, 126
McFall lease: 53
McGraw, J. J.: 123
McGuinn, Frank: 71
McIntosh County: 5
McIntyre, Edward F.: 123
Mackin, Jim: 18
McMan Oil Company: 122
McMillan farm: 45
Magee, Carl C.: 113
Magnolia Pipeline Company: 118, 135
Mannford North Field: 56
Maramec East Field: 47
Maramec Field: 47
Maramec North Field: 47
Maramec, Oklahoma: 115
Marietta, Ohio: 114
Mark K. Chapman Center for Communicative Disorders (Tulsa): 126
Marland No. 1 School Land Well: 43
Martin, L. J.: 112–113
Masham Field: 47; *see also* Donahue Field
Mellon, A. W.: 135
Mellon, W. L.: 81
Melton, W. A.: 123
Michaels farm: 53
Mid-Continent Building (Tulsa): 115
Mid-Continent Gas Company: 141
Mid-Continent-Gulf Coast Region: 72
Mid-Continent-Fort Arthur Pipe Line and Refining Company: 79
Mid-Continent Region: 55, 73, 89, 132
Midland Oil Refinery (El Dorado, Kansas): 123
Midland Valley Railroad: 87, 106, 118
Miller, Bluford: 16
Miller, "Shorty": 64–65
Milliken, John T.: 53
Milliken Oil Company: 53
Milliken, W. H.: 67, 70, 76, 106, 141
Minneapolis, Minnesota: 47
Minnehoma Field: 47
Minnehoma No. 2: 51
Minnehoma Oil and Gas Company: 55
Minehoma Oil Company: 47, 51, 53, 55
Minnesota: 51
Minnetonka Oil and Gas Company: 30
Mississippi Horizon: 75
Missouri: 3
Missouri, Kansas, and Texas Railroad: 34, 43, 81,108
MKT Railroad: *see* Missouri, Kansas and Texas Railroad
Missouri Medical College: 12
Mitchell, John O.: 66, 112
Moore, J. C.: 33
Moore, Zeke: 70
Morrison Field: 43; *see also* Watchorn Field

INDEX

Morrison, J. E.: 67
Mosier, M. H.: 70
Mounds, Oklahoma: 12, 68, 71, 87, 105–107, 140
Mounds Enterprise: 66
Mounds Field: 72; *see also* Glenn Pool
Mounds Sand: 75
Muskogean Nation: 3; *see also* Creek Nation
Muskogee County, Oklahoma: 5, 55
Muskogee, Oklahoma: 5, 14, 17, 67–68, 113–114
Muskogee Weekly Phoenix: 17

Nacogdoches, Texas: 81
Nathaniel Yargee No. 1: 22
Nathaniel Yargee No. 3: 62
Nation Carry: 11–112
National Bank of Tulsa: 119
National Guard in the oil fields: 85
National Petroleum News: 73
National Production Company: 137
Nederland, Texas: 132
Nellie Johnstone No. 1: 135
Nelson, Wilbur L.: 126
Neodesha, Kansas: 25, 120, 135
Neosho, Missouri: 115
Newblock, Herman: 113
New England Oil Company: 43
New Mexico: 132
Nickle Creek: 5
No Man's Land: 27
Nebraska: 28
Newkirk, Oklahoma: 27
Norman, Rena: 17
Norman, Sanny: 17

Oakley, Annie: 117
Oil City Derrick: 121
"Oil Flyer" special (train): 101
Oil & Gas Journal: 121
Oil Inspector, office of: 84
Oil Investors Journal, The: 121
Oil State Gasoline Company: 137
Okfushkee County, Oklahoma: 5
Oklahoma City: Oklahoma: 17, 22, 27, 55, 83, 113, 140
Oklahoma Corporation Commission: 85
Oklahoma County, Oklahoma: 56
Oklahoma Gas and Electric Company: 139
Oklahoma Gasoline Plant: 137
Oklahoma Historical Society: 10
Oklahoma-Kansas Division of the Mid-Continent Oil and Gas Association: 127
ONG: *see* Oklahoma Natural Gas Company
Oklahoma Natural Gas Company: 23, 83, 140
Oklahoma Natural Gas, Light and Heat Company: 27
Oklahoma Oil Company: 39
Oklahoma Pipeline Company: 118, 137
Oklahoma State Bank (Sapulpa): 88
Oklahoma State Oil Company: 114
Oklahoma Territory, oil in: 27–43
Okmulgee County, Oklahoma: 5
Okmulgee, Oklahoma: 12, 15, 81
Oliphant Hall (Tulsa): 126
Oliphant, John A.: 112
Oliphant, Mr. and Mrs. Allen G.: 126
Olney Field: 39
Omaha, Nebraska: 5
O'Neal, Jimmie: 22
Organic Act: 27
Osage and Oklahoma Company: 138
Osage and Oklahoma Gas Company: 83
Osage City, Oklahoma: 43
Osage County, Oklahoma: 43, 47, 53
Osage Inn (Osage City): 43
Osage Nation: 36, 47, 115
Osage Reservation: 27
Oswego Horizon: 75
Oto-Missouri Reservation: 27
Owen, Albert P.: 10
Ozark Pipeline Company: 135

Packaway, Ohio: 62
Page, Charles: 39, 45
Palm Beach, Florida: 115
Parkinson, James H.: 5
Pat Coyne Drugs (Tulsa): 108
Pawnee County, Oklahoma, oil in: 27–61
Pawnee East Field: 56
Pawnee Indians: 28
Pawnee Reservation: 27–28, 35
Payne County, Oklahoma: 43, 47, 56
Peck, Tom: 43
Pennsylvania: 87
PennWell Publishing Company: 122
Perry, Oklahoma: 62
Perryman Horizon: 75
Perryman, L. C.: 10
Pettee, William: 17
Petroleum Publishing Company: 122
Pilcade Building (Tulsa): 122
Phillips Engineering Building (Tulsa): 126
Phillips, Waite: 122, 126
Philtower Building (Tulsa): 122
Pierce Pipeline Company: 118, 135
Piersol, J. E.: 79
Pig's Ear (Tulsa): 111
Piqua Oil Company: 39
Pitman, Robert, Jr.: 69
Pittsburgh, Pennsylvania: 138
Poag, Oklahoma: 107
Pomeroy and Hamilton Oil Company: 47
Ponca Reservation: 27
Port Arthur, Texas: 76, 81, 132–133
Port Arthur-to-Humble, Texas pipeline: 133
Pleasant, Porter: 14, 17, 19, 23
Posey, Lena: 70, 105
Postoak, Lincoln: 17
Pottawatomie Reservation: 27
Prairie Oil and Gas Company: 22, 41, 70, 71, 76, 98, 106, 135
Prairie Pipeline Company: 118, 135
Praper: 94; *see also* Kiefer
Presbyterian Mission (Tulsa): 108
Price and Gillette General Store (Tulsa): 108
Price Field: 43
Price, P. L.: 113
Producers Oil Company: 133

Q County, Oklahoma Territory: 28
Quaker Oil Company: 76
Quay Field: 43
Quay North Field: 43
Queen, Emmet: 135

INDEX

Ralston, Oklahoma: 43
Ramona, Oklahoma: 135
Ranch Creek Field: 45
Randolph, H. W.: 113
Reavis, Holland S.: 121
Red Fork District: 3
Red fork Field: 3–26, 71
Red Fork Horizon: 75
Red Fork, Oklahoma: 3, 17–27, 62, 108, 117–120
Red Fork Sand: 39
Red River: 81
Reed, J. H.: 135
Republic Oil and Pipeline Company: 45, 47
Rhodes, John: 70, 106
Robinson, Colonel: 17
Robinson, Edward: 17
Robinson, George: 17
Robinson Hotel (Tulsa): 109, 111
Robinson, William H.: 109
Roddy, F. M.: 113
Roser, Mrs. Charles: 115
Roser, W. B. (Billy): 114–115
Rose, D. C.: 79
Roosevelt, Theodore: 84
Rucker, Matt: 106
"Rural Courtship" (painting): 141
Rusk County, Texas: 81
Russell farm: 53

Sabo, Oklahoma: 87, 105
Sac and Fox Reservation: 27
Saint James Hotel (Sapulpa): 88
St. Elmo Hotel (Tulsa): 108
St. Louis, Missouri: 5, 34, 89, 108, 115
St. Louis and San Francisco Railroad: 5, 9–10, 17, 79, 81, 87–88, 94, 105–106, 108, 119
Sand Springs, Oklahoma: 3
Saratoga, Texas: 133
Sanderson, C. A.: 113
Sanger, Lena: 17
Sanger, May: 17
Sapulpa, Oklahoma: 3, 10, 63–64, 72, 83
Sapulpa Electric Company: 91
Sapulpa and Interurban Railway: 88
Sapulpa Ice Company: 88
Sapulpa Field: 72; *see also* Glenn Pool
Sapulpa, James: 10, 87
Sapulpa, Oklahoma: 87–88, 91, 94, 104–105, 107
Sapulpa State Bank: 88
Sapulpa, William A.: 10
Sawyer and Company: 76
Sawyer, C. J.: 113
Schell, John: 30, 32–33
Schoenfelt Oil Company: 39
Scioto Oil Company: 39
Selby, J. M.: 70
Selby Oil & Gas Company: 43, 76
Self, Buck: 106
Seminole County, Oklahoma: 5, 56
Seminole Field: 55, 142
Seminole Nation: 5
Severs, F. B.: 12, 15
Shaffer Pipeline Company: 135
Shawnee, Oklahoma: 27, 62, 83
Shawnee Oil Company: 76
Shobe, Charlie: 96–98, 104
Simley E. L.: 10
Sinclair, Earl W.: 73
Sinclair, Harry F.: 73, 81, 91, 109, 118
Sinclair Oil and Gas Company: 73–74, 91
Sinclair Oil and Refining Company: 73
Sinclair Pipeline Company: 118, 135
Skedee Field: 56
Skedee North Field: 56
Skedee West Field: 56
Skelly Oil Company: 122–123
Skelly Stadium (Tulsa): 126
Skelly-Sankey Oil Company: 123
Skelly, William G.: 122–124, 126
Skinner Sand: 39
Smith County, Texas: 81
Sneed, Earl: 123
Society of Exploration Geophysicists: 127
Sour Lake, Texas: 81, 133
South Cleveland Field: 39
Southland Gasoline Company: 137
South McAlester, Oklahoma: 81
Southwestern Association of Petroleum Geologists: 127
Southwestern Field: 47
Spence Kellum Livery Stable (Tulsa): 108
Spindletop Field: 9, 27, 132–133
Standard Oil Company: 71, 81
Standard Oil of New Jersey: 41
State Tad Oil Company: 39
Stickles, Mrs. Vera Bland: 14
Stout's Ferry (Cleveland): 43
Stryker, William: 79
Sue A. Bland No. 1: 10–26, 62, 117
Sulzbacher, T. C.: 69
Sun Company: 121
Sun Petroleum Products Company: 121
Sunny Slope Southwest Gas Field: 56
Superior Oil and Gas Company: 18
Sutton, G. W.: 34
Sylvan Horizon: 75

Tallahasee: 3, 108; *see also* Tulsa
Tallasi: 108, *see also* Tulsa
Taneha, Oklahoma: 87, 105
Taneha Horizon: 75
Taneha Field: 83
Taylor, Charles H.: 126–127
Taylor, Nancy: 55
Templeton, Walter E.: 53
Terlton Field: 43
Terlton North Field: 43
Terlton, Oklahoma: 43
Terlton Southeast Pool: 43
Terre Haute, Indiana: 117
"Tex" (saloonkeeper): 104
Texaco: *see* Texas Company, The
Texas: 3, 9, 72, 132
Texas Barge No. 1: 133
Texas Company, The: 79, 81, 105, 120, 132–133, 135, 137
Texas Pipline Company: 118, 135
Thomas, J. Elmer: 126
Thomas Gilcrease Institute of American History and Art (Tulsa): 142
Thomas, Usher: 17
Three Forks: 3
Tide Water Oil Company: 67,73
Titusville, Pennsylvania: 5

INDEX

Tonkawa Reservation: 27–28
Transcontinental Oil Company: 137
Tucker, H. H.: 79
Tucker Sand: 36
Tuloma Oil Company: 141
Tulsa Commercial Club: 26, 79, 111–112, 117, 120, 125
Tulsa Banking Company: 108
Tulsa County, Oklahoma: 3, 5, 83
Tulsa Democrat: 17, 66
Tulsa Field: 72, 108; *see also* Glenn Pool
Tulsa, Oklahoma: 3, 14, 17–18, 26, 34, 71, 73, 81, 84, 88, 101, 107–132, 138–141
Tulsa World: 108
Tulsey Town: 3, 108; *see* Tulsa
Turkey Creek: 62
Turkey Mountain Horizon: 75
Tyler, Texas: 81
Tyrrell Hall (Tulsa): 126
Tyrrell, Harry C.: 113, 126

Ufer, F. B.: 118
Uncle Bill No. 1: 30, 32–34, 43, 56–57
Uncle Sam Oil Company: 45
Uncle Sam Refinery (West Tulsa): 79, 120
University of Oklahoma, Geological Department: 126
University of Tulsa: School of Petroleum Engineering, 125; School of Petroleum Geology: 126
Unassigned Lands: 62
Upper Creek: 3
Upshaw, Taz: 17

Valley Osage Oil Company: 138
Vanderver, W. A.: 123
Van Winkle, Lee: 17
Verdigris River: 3
Vowell, Hazel: 106

Wagoner County, Oklahoma: 5
Waibel, Ted: 87
Walker, P. G.: 53
Wallace and Calhoun Meat Market (Tulsa): 108
Warren, F. L.: 53
Warren, William K.: 124
Washington County, Oklahoma: 73
Washington, D. C.: 117
Watchorn East Field: 45
Watchorn Field: 43, 45; *see also* Morrison Field
Watchorn North Field: 45
Watchorn Northwest Field: 45
Watchorn West Field: 45
Water-Pierce Oil Company: 121
Watkins Camp: 107
Watkins, Oklahoma: 81
West Quay Field: 43
West Tulsa: 115, 120–121
West Virginia: 87
White, Pat J.: 118
Whiting, Indiana: 71, 135
Wichita and Caddo Reservation: 27–28
Wick and Company: 11
Wick, John S.: 10–26
Wilcox Horizon: 75
Wilcox Sand: 75
Williams Companies, The. 122
Williams, David R.: 122
Williams S. Miller: 122
Williamson Brothers store (Tulsa): 108
Williamson, George T.: 26, 120
Wilson, Boston: 17
Wilson, J. B.: 113
Winkler Harness (Tulsa): 108
Wirt, Oklahoma: 116
Wood County, Texas: 81
Wooden, John N.: 113
W. O. Struther Drugs (Tulsa): 108
Wright, J. George: 19, 23
Wright, J. H.: 138

Yargee, John: 32
Yargee, John I.: 10
Yargee, Mrs. John: 22
Yargee, Mr. and Mrs. John I.: 17
Yargee, Pleas: 17
Yargee, Van: 17, 22

Zoldoski No. 1: 47